ADVANCED LEARNER'S DICTIONARY *Of* BIOLOGY

ADVANCED LEARNER'S DICTIONARY

Of

BIOLOGY

Compiled & Edited by

A Team of Experts

ANMOL PUBLICATIONS PVT. LTD.

NEW DELHI - 110 002 (INDIA)

ANMOL PUBLICATIONS PVT. LTD.
4374/4B, Ansari Road, Daryaganj
New Delhi - 110 002

Advanced Learner's Dictionary of Biology

First Edition, 2000
Reprint 2001
ISBN 81-261-0465-1

[Responsibility for the facts stated, opinions expressed, conclusions reached and plagiarism, if any, in these volumes is entirely that of the Editor. The Publishers bear no responsibility for them whatsoever.]

PRINTED IN INDIA

Published by J.L. Kumar for Anmol Publications Pvt. Ltd., New Delhi and Printed at Mehra Offset Press, Delhi.

Preface

This *Advanced Learner's Dictionary of Biology* is basically designed for students and teachers as well. It is the result of a careful analysis of the needs of advanced level students and should be an invaluable aid in their day-to-day learning.

As it is extremely difficult to limit the contents of such a dictionary at 'Advanced' level standard, many headwords are included that are above this level and that may be of interest to Advanced Learners. A few headwords that are not strictly scientific are also included because of their common usage in everyday scientific terminology. It has been our aim to provide explanations that are easily understood, whilst maintaining academic standards.

We have tried to include the more important terms used at advanced level, but would welcome any comments regarding serious omissions or indeed about the content of any of the entries.

We would like to thank all the people who have cooperated in producing this dictionary. We are also grateful to many people who have given additional help and advice.

—Editors

A

ABO system: One of the most important human blood group systems. The system is based on the presence or absence of antigens A and B on the surface of red blood cells and antibodies against these in blood serum. A person whose blood contains either or both these antibodies cannot receive a transfusion of blond containing the corresponding antigens as this would cause the red cells to clump. The table illustrates the basis of the system: people of blood group O are described as 'universal donors' as they can give blood to those of any of the other groups.

Abscisic acid: A naturally occurring plant growth substance that promotes leaf ageing, leaf fall, and apical dominance and induces dormancy in seeds and buds.

Abscission: The separation of a leaf, fruit, or other part front the body of a plant. The process is controlled by growth substances, notably abscisic acid; it involves the formation of an abscission *zone,* at the base of the part> within which a layer of cells (*abscission layer)* breaks down.

Absorption: The movement of fluid or a dissolved substance across a cell membrane. In animals, for example, soluble food material is absorbed into the circulatory system, through cells lining the alimentary canal. In plants, water and mineral salts are absorbed front the soil by the roots.

Abyssal zone: The lower depths of the ocean (below approximately 2000 metres), where there is effectively no light penetration. Abyssal organisms are adapted for living under high pressures in cold dark conditions.

Accommodation: (in animal physiology) The process by which the focal length of the lens of the eye is changed so that clear images of objects at a range of distances arc displayed on the retina, In man and sortie other mammals accommodation is achieved by reflex adjustments in the shape of the lens brought about by relaxation and contraction of muscles within the ciliary body. 2. (in botany) The ability of a plant to adapt itself to gradually changing environmental conditions. 3. (in animal behaviour) The psychological adjustment made by an animal in response to continuously changing environmental conditions.

Acellular: Describing tissues or organisms that are not made up of separate cells but often have more than one nucleus.

Acetic acid (ethanoic acid): A carboxylic acid, CH_3COOH, that, when combined with coenzyme A (to form *acetyl coenzyme* A), plays a crucial role in energy metabolism.

Acetylcholine: A substance that is released at some (*cholinergic*) nerve endings. Its function is to pass on a nerve impulse to the next nerve (i.e., at a synapse) or to initiate muscular contraction. Once acetylcholine has been released it has only a transitory effect because it is rapidly broken down by the enzyme *acetylcholinesterase*,

Achene: A dry indehiscent fruit formed from a single carpel and containing a single seed. An example is the feathery achene of clematis.

Acid anhydride: A type of organic compound of general formula RCOOCO'R, where R and R' are alkyl or aryl groups. They are prepared by reaction of an acyl halide with the sodium salt of a carboxylic acid.

Acid-base balance: The regulation of the concentrations of acids and bases in blood and other body fluids such that the p^H remains within a physiologically acceptable range. This is achieved by the presence of natural buffet systems, such as the haemoglobin, bicarbonate ions, and carbonic acid in mammalian blond. By acting in conjunction, these effectively mop up, excess acids and bases and therefore prevent any large shifts in blood p^H. The acid-base balance is also influenced by the selective removal of certain ions by the kidneys and the rate of removal of carbon dioxide from the lungs.

Acquired characteristics: Features that are developed during the lifetime of an individual, e.g. the enlarged arm muscles of a tennis player. It is a basic tenet of current evolutionary thought that such characteristics are not genetically controlled and cannot be passed on to the next generation.

ACTH (adrenocorticotrophic hormone, corticotrophin): A hormone produced by the anterior Pituitary gland in response to stress that controls secretion of certain hormones (the corticosteroids) by the adrenal glands. It can be administered by injection to treat such disorders as rheumatic diseases and asthma, but it only relieves symptoms and is not a cure.

Actinomycetes: A group of Gram-positive mostly anaerobic nonmotile bacteria. All species are fungus-like, with filamentous cells producing reproductive spores on aerial branches similar to the spores of certain moulds, The group includes bacteria of the genera *Actinomyces*, some species of which cause disease in animals and man; and *Streptomyces*, which are a source of many important antibiotics (including streptomycin).

Action potential: The change in electrical potential that occurs across a cell membrane during the passage of a nerve impulse, As an impulse travels in a wavelike manner along the axon of a nerve, it causes a localized and transient switch in electrical potential across the cell membrane from 60 mV (millivolts) to +45 mV. Nervous stimulation of a muscle fibre has a similar effect.

Action spectrum: A graphical plot of the efficiency of clectromagnetic radiation in producing a photochemical reaction against the wavelength of the radiation used, For example, the action spectrum for photosynthesis using light shows a peak in the region 670-700 nm. This corresponds to a maximum absorption in the absorption spectrum of chlorophylls in this region.

Active site: The site on the surface of an enzyme molecule that binds the substrate molecule. The properties of an active site are determined by the three-dimensional arrangement of the polypeptide chains of the enzyme and their constituent amino acids. These govern the nature of the interaction that takes place and hence the degree of substrate specificity and susceptibility to inhibition.

Active transport: The movement of substances through membranes in living cells' often against a concentration gradient: a process requiring metabolic energy. Organic molecules and inorganic ions are transported into and out of both cells and their organelles. It is thought that the substance binds to a carrier protein embedded in the membrane, which carries it through the membrane and releases it on the opposite side. Active transport serves chiefly to maintain the normal balance of ions in colts, especially the concentration gradients of sodium and potassium ions crucial to the activity of nerve and muscle cells.

Adaptation: (in evolution) Any change In the structure or functioning of an organism that makes it better suited to its environment. Natural selection of inheritable adaptations ultimately leads to the development of new species. Increasing adaptation of a species to a particular environment tends to diminish its ability to adapt to any sudden change in that environment. (in physiology) The alteration in the degree of sensitivity (either an increase or a decrease) of a sense organ to suit conditions more extreme than normally encountered.

Adaptive radiation (divergent evolution): The evolution from one species of animals or plants of a number of different forms. As the original population increases in size it spreads out front its centre of origin to exploit new habitats and food sources. In time this results in a number of populations each adapted to its particular habitat: eventually these populations will differ front each other sufficiently to become new species. A good example of this process is the evolution of the Australian marsupials into species adapted as carnivores, herbivores, burrowers, fliers, etc. On a smaller scale, the adaptive radiation of the Galapagos finches provided Darwin with crucial evidence for his theory of evolution.

Adenosine: A nucleoside comprising one adenine molecule linked to a D-ribose sugar molecule. The phosphate-ester derivatives of adenosine, AMP, ADP, and ATP, are of fundamental biological importance as carriers of chemical energy.

Adenovirus: One of a group of DNA-containing viruses found in rodents, fowl, cattle, monkeys, and man, In man they produce acute respiratory-tract infections with symptoms resembling those of the common cold. They are also implicated in the formation of tumours.

Adipose tissue: A body tissue comprising cells containing fat and oil. It is found chiefly below the skin and around major organs (such as the kidneys and heart), acting as an energy reserve and also providing insulation and protection.

Adrenal glands: A pair of endocrine glands situated immediately above the kidneys (hence they are also known as the suprarenal glands). The inner portion of the adrenals, the medulla, secretes the hormones adrenaline and noradrenaline; the outer cortex secretes small amounts of sex hormones (androgens and oestrogens) and various corticosteroids, which have a wide range of effects on the body.

Adrenaline (epinephrine): A hormone, produced by the medulla of the adrenal glands, that increases heart activity, improves the power and problems the action of muscles, and increases the rate and depth of breathing to prepare the body for 'fright, flight, or fight'. At the same time it inhibits digestion and excretion. Similar effects are produced by stimulation of the sympathetic nervous system Adrenaline can be administered by injection to relieve bronchial asthma and reduce blood loss during surgery constricting blood vessels.

Adrenergic Describing a nerve fibre that either releases adrenaline or noradrenaline when stimulated or is itself stimulated by these substances.

Adsorption: The formation of a layer of solid, liquid, or gas on the surface of a solid or, less frequently, of a liquid. There art two types depending on the nature of the forces involved, In *chemisorption a* single layer of molecules, atoms, or ions is attached to the adsorbent surface by chemical bonds. In *physisorption* adsorbed molecules are held by the weaker physical forces. The property is utilized in adsorption chromatography.

Aestivation: (in zoology) A state of inactivity occurring in some animals, notably during prolonged periods of drought or heat. Feeding, respiration, movement, and other bodily activities are considerably slowed down. (in botany) The arrangement of the parts of a flower bud, especially of the sepals and petals.

Afferent: Carrying (nerve impulses, blood, etc.) from the outer regions of a body or organ towards its centre. The term is usually applied to types of nerve fibres or blood vessels.

Aflatoxin: Any of four related toxic compounds produced by the mould *Aspergillus flavus*. Aflatoxins bind to DNA and prevent replication and transcription. They can muse acute liver damage and cancers: man may be poisoned by eating stored peanuts and cereals contaminated with the mould.

Afterbirth: The placenta umbilical cord, and extraembryonic membranes, which are expelled from the womb after a mammalian fetus is born. In most nonhuman mammals the afterbirth, which contains nutrients and might otherwise attract predators, is eaten by the female.

Agar: An extract of certain species of red seaweeds that is used as a gelling agent in microbiological culture media, foodstuffs, medicines, and cosmetic creams and jellies. *Nutrient agar* consists of a broth made from beef extract or blood that is gelled with agar and used for the cultivation of bacteria, fungi, and some algae.

Aggression: Bebaviour aimed at intimidating or injuring another animal of the same or a competing species. Aggression between individuals of the same species often starts with a series of ritualized displays or contests that car, end at any stage if one of the combatants withdraws, leaving the victor with access to a disputed resource (e g.-food, a mate, or territory) or with increased social dominance. It is also often seen in courtship. Aggression or threat displays usually appear to exaggerate performer's size or strength; for example, many fish erect their fins and mammals and birds may erect hairs or feathers. Special markings may be prominently exhibited, and in*tention* movements may be made: dogs bare their teeth, for example.

Agnatha: A class of marine and freshwater vertebrates that lack jaws. They are fishlike animals with cartilaginous skeletons and well-developed sucking mouthparts with horny teeth, The only living agnathans are lampreys and hagfishes (order Cyclostomata), which are parasites or scavengers, Fossil agnathans, covered in an armour of bony plates, are the oldest known fossil vertebrates. They have been dated from the Silurian and Devonian periods, 440-345 million years ago.

Albinism: Hereditary lack of pigmentation in an organism. Albino animals and human beings have no colour in their skin, hair, or eyes. The allele responsible is recessive to the allele for normal pigmentation.

Albumin (albumen): one of a group of globular proteins that are soluble in water but form insoluble coagulates when heated. Albumins occur in egg white, blood, milk, and plants. Serum albumins, which constitute about 55% of blood plasma protein, help regulate the osmotic pressure and hence plasma volume. They also bind and transport fatty acids. a-lactalbumin is one of the proteins in milk.

Aldosterone: A hormone produced by the adrenal glands that controls excretion of sodium by the kidneys and thereby maintains the balance of salt and water in the body fluids.

Algae: A large and diverse group of simple plants that contain chlorophyll (and can therefore carry out photosynthesis) and live in aquatic habitats and in moist situations on !and The plant body may be unicellular or multicellular (filamentous, ribbon-like, or platelike). The algae are not now regarded as a taxonomic group, although they are sometimes grouped with other simple organisms in the kingdom Protista. Usually, however, the component groups are, recognized as taxonomic divisions. Separation into these divisions is based primarily on the composition of the cell wall, the nature of the stored food reserves, and the other photosynthetic pigments present.

Alimentary canal (digestive tract; gut): A tubular organ in animals that is divided into a series of zones specialized for the ingestion, digestion, and absorption of food and for the elimination of indigestible material. In most animals the canal has two openings, the mouth (for the intake of food) and the anus (for the elimination of waste). Simple animals, such as coelenterates (e.g. *Hydra* and jellyfish) and flatworms, have only one opening to their alimentary canal, which must serve both functions.

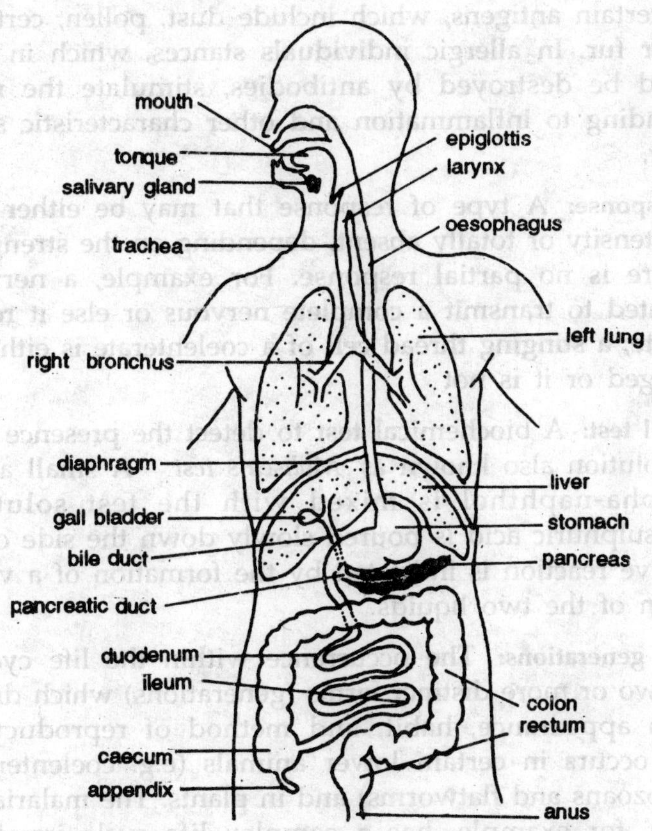

The alimentary canal

Alkaloid: One of a group of nitrogenous organic compounds derived front plants and having diverse pharmacological properties. Alkaloids include morphine, cocaine, atropine, quinine, and caffeine, most of which are used in medicine as analgesics (pain relievers) or anaesthetics. Some alkaloids are poisonous, e.g., strychnine and coniine, and colchicine inhibits cell division.

Allantois: One of the membranes that develops in embryonic reptiles, birds, and mammals as a growth from the hindgut. It acts as a urinary bladder for the storage of waste excretory products in the egg (in reptiles and birds) and as a means of providing the embryo with oxygen (in reptiles, birds, and mammals) and food.

Allergy: A condition in which the body produces an abnormal immune response to certain antigens, which include dust, pollen, certain foods and drugs, or fur. In allergic individuals stances, which in a normal person would be destroyed by antibodies, stimulate the release of histamine, leading to inflammation and other characteristic symptoms of the allergy.

Ali-or-none response: A type of response that may be either complete and of full intensity or totally absent, depending on the strength of the stimulus; there is no partial response. For example, a nerve cell is either stimulated to transmit a complete nervous or else it remains in its resting state; a stinging thread cell of a coelenterate is either Completely discharged or it is not

Alpha-naphthol test: A biochemical test to detect the presence of carbohydrates in solution also known as *Molisch's test* A small amount of alcoholic alpha-naphthol is mixed with the test solution and concentrated sulphuric acid is poured slowly down the side of the test tube. A positive reaction is indicated by the formation of a violet ring at the junction of the two liquids.

Alternation of generations: The occurrence within the life cycle of an organism of two or more distinct forms (generations) which differ from each other in appearance, habit, and method of reproduction. The phenomenon occurs in certain lower animals (e.g. coelenterates and parasitic protozoans and flatworms) and in plants. The malaria parasite (Plasmodium), for example, has a complex life cycle involving the alternation of sexually and asexually reproducing generations. In plants the generation with sexual reproduction is called the gametophyte and the asexual generation is the sporophyte, either of which may dominate the life cycle, and there is also alternation of the haploid and diploid states. Thus in ferns the dominant plant is the diploid sporophyte; it produces spores that germinate into small haploid gametophytes bearing sex organs. In mosses the gametophyte is the dominant plant and the sporophyte is the spore-bearing capsules.

Altruism: Behaviour by an animal that decreases its chances of survival or reproduction while increasing those of another member of the same species. For example, a lapwing puts itself at risk by luring a predator away from the nest through feigning injury, but by so doing saves its offspring, Altruism in its biological sense does not imply any conscious benevolence on the part of the performer. Altruism can evolve through kin selection, if the recipients of altruistic acts tend on average to be more closely related to the altruist than the population as a whole.

Alveolus: The tiny air sac in the lung of mammals and reptiles at the end of each bronchiole. It is lined by a delicate moist membrane, has many blood capillaries, and is the site of exchange of respiratory gases (carbon dioxide and oxygen). 2, The socket in the jawbone in which a tooth is rooted by means of the periodontal membrane.

Amber: A yellow or reddish-brown fossil resin. The resin was exuded by certain trees and other plants and often contains preserved insects, flowers, or leaves that were trapped by its sickly surface before the resin hardened. Amber is used for jewellery and ornaments. It also, has the property of acquiring an electrical charge when rubbed. It occurs throughout the world in rock strata from the Cretaceous to the Pleistocene, but most commonly in Cretaceous and Tertiary rocks.

Amino acid: Any of a group of water-soluble organic compounds that possess both a carboxyl (-COOH) and an amino (-NH_2) group attached to the a-carbon atom Amino acids can be represented by the general formula R-CH(NH_2)COOH. R may be hydrogen or an organic group and determines the properties of any particular amino acid. Through the formation of peptide bonds, amino acids join together to form short chains (peptides) or much longer chains (polypeptides). Proteins arc coin posed of various proportions of about 20 commonly occurring amino acids. The sequence of these amino acids in the protein polypeptides determines the shape, properties, and hence biological role of the protein. Some amino acids that never occur in proteins are nevertheless important, e.g. ornithine and citrulline, which are intermediates in the urea cycle.

Plants and many microorganisms can synthesize amino acids front simple inorganic compounds, but animals rely on adequate supplies in their diet. The essential amino acids must be present in the diet whereas others can be manufactured from them.

Ammonite: An extinct aquatic mollusc of the class Cephalopoda. Ammonites were abundant in the Mesozoic era (225-65 million years ago) and arc commonly found as fossils in rock strata of that time, being used as index fossils for the Jurassic period. They were characterized by a coiled shell divided into many chambers, which acted as a buoyancy aid. The external suture lines on these shells increased in complexity with the advance of die group.

Amniocentesis: The taking of a sample of amniotic fluid front a pregnant woman to determine the condition of an unborn baby. A hollow needle is inserted through the woman's abdomen and wall of the uterus and the fluid drawn off. Chemical and microscopical examination of cells shed from the embryo's into the fluid arc used to detect spina bifida, Down's syndrome, or other serious biochemical or chromosomal abnormalities.

Amnion: A membrane that encloses the embryo of reptiles, birds, and mammals within the amniotic cavity. This cavity is filled with amniotic fluid in which the embryo is protected front desiccation and from external pressure.

Amniote: A vertebrate whose embryos are totally enclosed in a fluid-filled sac the amnion. The evolution of the amnion provided the necessary fluid environment for the developing embryo and therefore allowed animals to breed away front water. Amniotes comprise the reptiles, birds, and mammals.

Amoeba: A genus of Protozoa, members of which have temporary body projections called pseudopodia. These are used for locomotion and feeding and result in a constantly changing body shape. Most species are free-living in soil, mud, or water, where they feed on smaller protozoans and single-celled plants, but a few are parasitic.

Amoebocyte: An animal cell whose location is not fixed and is therefore able to wander through the body tissues. Amoebocytes are named after their resemblance especially, in their movement, to Amoeba and they feed on foreign particles.

Amount of substance: Symbol *n*. A measure of the number of entities present in a substance. The specified entity may be an atom, molecule, ion, electron, photon, etc., or any specified group of such entities. The amount of substance of an element, for example, is proportional to the number of atoms present. The SI unit of amount of substance is the mole.

Amphibia: The class of vertebrates that contains the frogs, toads, newts, and salamanders. Amphibians evolved in the Devonian period as the first vertebrates to occupy the land, and many of their characteristics are adaptations to terrestrial life. All adult amphibians have a passage linking the roof of the mouth with the nostrils so they may breathe air and keep the mouth closed. The moist scaleless skin is used to supplement the lungs in gas exchange. They have no diaphragm, and therefore the muscles of the mouth and pharynx provide the pumping action for breathing. Fertilization is usually external and the eggs are soft and prone to desiccation, therefore reproduction commonly occurs in water. Amphibian larvae are aquatic, having gills for respiration; they undergo metamorphosis to the adult form.

Amylase (diastase): Any of a group of closely related enzymes that degrade starch, glycogen, and other polysaccharides. Plants contain both α and β–amylases; animals possess only α-amylases, found in pancreatic juice and also (in humans and some other species) in saliva. Amylases cleave the long polysaccharide chains, producing a mixture of glucose and maltose.

Amylose: A polysaccharide consisting of linear chains of between 100 and 1000 linked glucose molecules. Amylose is a constituent of starch In water, amylose reacts with iodine to give a characteristic blue colour.

Anabolic steroid: Any steroid compound that promotes tissue growth, especially of muscles. Naturally occurring anabolic steroids include the male sex hormones. Synthetic forms of these are used medically to help weight gain alter debilitating diseases; their use by athletes to build up body muscles can cause liver damage and is banned by most athletic authorities.

Anabolism: The metabolic synthesis of proteins, fats, and other constituents of living organisms front molecules or simple precursors. This process requires energy in the form of ATP.

Anaerobic respiration: A type of respiration in which foodstuffs (usually carbohydrates) are partially oxidized, with the release of chemical energy, in a process not involving atmospheric oxygen. Since the substrate is never completely oxidized the energy yield of this type of respiration is lower than that of aerobic respiration. It occurs in some yeasts and bacteria and in muscle tissue when oxygen is absent. *Obligate anaerobes* are organisms that cannot use free oxygen for respiration; *facultative anaerobes* are normally aerobic but can respire anacrobically during periods of oxygen shortage.

Analogous: Describing features of organisms that are superficially similar but have evolved in different ways. The wings of butterflies and birds are analogous organs.

Anamniote: A vertebrate that lacks an amnion and whose embryos and larva must therefore develop in water. Anamniotes comprise the agnathans, fishes, and amphibians.

Anaphase: The third stage of cell division. In mitosis the chromatids of each chromosome move apart to opposite ends of the spindle. In the first anaphase of meiosis, the paired homologous chromosomes separate and move to opposite ends; in the second anaphase the chromatids move apart, as in mitosis.

Anaphylaxis: An abnormal immune response that occurs when an individual previously exposed to a particular antigen is re-exposed to the same antigen. Anaphylaxis may follow an insect bite or the injection of a drug (such as penicillin). It is caused by the release of histamine and similar substances and may produce a localized reaction or a more generalized and severe one, with difficulty in breathing, pallor, or drop in blood pressure, unconsciousness, and possibly heart failure and death.

Anatomy: The study of the structure of living organisms, especially of their internal parts by means of dissection and microscopical examination.

Androgen: One of a group of male sex hormones that stimulate development of the testes and of male secondary sexual characteristics (such as growth of facial and pubic hair in men). *Testosterone is* the most important. Androgens are produced principally by the testes when stimulated with luteinizing hormone but they are also secreted in smaller amounts by the adrenal glands and the ovaries. Injections of natural or synthetic androgens are used to treat hormonal disorders of the testes and breast cancer and to build up body tissue.

Anemophily: Pollination of a flower in which the pollen is carried by the wind. Examples of anemophilous flowers are those of grasses and conifers.

Angiospermae: The flowering plants: a subdivision of the Spermatophyta or a class of the Pteropsida. Angiosperm gametes are produced within flowers and the ovules (and the seeds into which they develop) are enclosed in a carpel. The angiosperms are the dominant plant forms of the present day. They show the most advanced structural organization in the plant *kingdom,* enabling them to inhabit a very diverse range of habitats. There are two classes (or subclasses) within this group: the Monocotyledonae with one seed leaf (cotyledon) in the seed, and the Dicotyledonae with two seed leaves.

Angiotensin: Either of two related peptide hormones that raise blood pressure. Angiotensin I is derived from a protein *(angiotensinogen)* secreted by the liver into the bloodstream. As blood passes through the lungs, an enzyme splits angiotensin I, forming angiotensin II. This causes constriction of blood vessels and stimulates the release of the hormones vasopressin and aldosterone, which increase blood pressure.

Angstrom: Symbol A. A unit of length equal to 10^{-10} metre. It was formerly used to measure wavelengths and inter-molecular distances but has now been replaced by the nanometre. 1A 0.1 nanometre. The unit is named after the Swedish pioneer of spectroscopy A. J. Angstrom.

Animal: Any living organism of the kingdom Animalia, characterized by an inability to manufacture its own food, so that it feeds on other organisms or organic matter (holozoic nutrition), Animals are therefore typically mobile (to search for food) and have evolved specialized sense organs for detecting changes in the environment; a nervous system coordinates information received by the sense organs and enables rapid responses to, environmental stimuli. Animal cells lack the cellulose cells walls of plant cells.

Animal behaviour: The activities that constitute an animal's response to its external environment. Certain categories of behaviour are seen in all animals (e.g. feeding, reproduction) but these activities involve different movements in different species and develop in different ways. Some movements are highly characteristic of a species *(see* instinct) whereas others are more variable and depend or, the interaction between innate tendencies and learning during the individual's lifetime. Physiologists study how changes in the body (e.g. hormone levels) affect behaviour, psychologists study the mechanisms of learning, and ethologists study the behaviour of the whole animal how this develops during the individual's lifetime and how it evolved through natural selection.

Annelida: A phylum of invertebrates corn prising the segmented worms (e.g. the earthworm), Annelids have cylindrical soft bodies showing metameric segmentation, obvious externally as a series of rings separating the segments. Each segment is internally separated front the next by a membrane and bears stiff bristles. Between the gut and other body organs there is a fluid-filled cavity called the coelom, which acts as a hydrostatic skeleton. Movement is by alternate contraction of circular and longitudinal muscles in the body wall. The phylum contains three classes: Polychaeta, Oligochaeta, and "Hirudinea.

Annulus: (in botany) a. A ragged ring of tissue that remains on the stalk of a mushroom or toadstool. Also called a *velum,* it is formed front the ruptured membrane that originally covered the lower surface of the cap, b. The region of the wall of a fern sporangium that is specialized for spore dispersal. It consists of cells that are thickened except on their outer walls. On drying out, the cells contract and the sporangium ruptures, releasing the spores. The annulus springs back into position when the residual water in the cells vaporizes and any remaining spores are dispersed. 2. (in zoology) Any of various ring-shaped structures in animals, such as any of the segments of an earthwortn or other annelid.

Antagonism: The interaction of two sub stances (e.g. drugs, hormones, or enzymes) having opposing effects in a system in such a way that the action of one partially or completely inhibits the effects of the other, For example, one group of anti-cancer drugs acts by antagonizing the effects of certain enzymes controlling the activities of the cancer cells. An interaction between two muscles in which contraction of one (the *antagonist)* prevents that of the other (the *agonist).* The antagonist must relax to enable the agonist to contract and effect movement. 1 An interaction between two organisms (e.g. moulds or bacteria) in which the growth of one is inhibited by the ether.

Antenna: A long whiplike jointed mobile paired appendage on the head of many arthropods, usually concerned with the senses of smell, touch, etc. In insects, millipedes, and centipedes they are the first pair of head appendages and are specialized and modified in marry insects. In crustaceans they are the second pair of head appendages, the first pair (the *antennules)* having the sensory function, while the antennae are modified for swimming and for attachment.

Anterior: 1. Designating the part of an animal that faces to the front, i.e., that leads when the animal is moving. In man and bipedal animals the anterior surface corresponds to the ventral surface. 2. Designating the side of a flower or auxiliary bud that faces away front the flower stalk or main stern, respectively.

Antheridium: The male sex organ of algae, fungi, bryophytes, and pteridophytes. It produces the male gametes *(antherozoids).* It may consist of a single cell or it may have a wall that is made up of one or several layers forming a sterile jacket around the developing gametes.

Antherozoid (spermatozoid): The motile male gamete of algae, fungi bryophytes, pteridophytes, and certain gymnosperms. Antherozoids usually develop in an antheridium but in certain gymnosperms, such as Ginkgo and Cycas, they develop front a cell in the pollen tube.

Anthocyanin: One of a group of flavonoid pigments. Anthocyanins occur in the cell vacuoles of various plant organs and are responsible for many of the blue, red, and purple colours in plants (particularly in flowers).

Antibiotics: Substances obtained front microorganisms, especially moulds, that destroy or inhibit the growth of other microorganisms, particularly disease-producing bacteria and fungi. Common antibiotics include penicillin, streptomycin, and tetracyclines.

Antibody: A protein produced by certain white blood cells (lymphocytes) in response to entry into the body of a foreign substance (antigen) in order to render it harmless. An antibody-antigen reaction is highly specific. Antibody production is one aspect of the immune response, and is stimulated by such antigens as invading bacteria, foreign red blood cells, inhaled pollen grains or dust, and foreign tissue grafts.

Anticoagulant A substance that prevents the formation of blood clots. Heparin is a natural anticoagulant, which is extracted to treat such conditions as thrombosis and embolism.

Antigen: Any substance that the body regards as foreign and that therefore elicits an immune response. Antigen may be formed in, or introduced into, the body. They are usually proteins. *Transplantation* antigens are associated with the tissues and are involved in tissue or organ grafts; an example is the system. A graft will be rejected if the recipient's body regards such antigens on the donor's tissues as foreign.

Antihistamine: Any drug that inhibits the effects of histamine in the body and is therefore used to relieve and prevent the symptoms associated with allergic reactions, such as hay fever.

Antiseptic Any substance that kills or inhibits the growth of disease-causing micro-organisms but is essentially nontoxic to cells of the body. Common antiseptics include hydrogen peroxide, the detergent cetrimide, and ethanol.

Aorta: The major blood vessel in higher vertebrates through which oxygenated blood leaves the heart from the left ventricle. The aorta branches to form many smaller arteries, which in turn branch many times to supply oxygen and essential nutrients to all living cells in the body.

Apical dominance: Inhibition of the growth of lateral buds in a plant by the presence of a growing apical bud. It is brought about by the action of auxins (produced by the apical bud) and abscisic acid.

Apical meristem: A region at the tip of each shoot and root divisions are continually occurring to produce new stem and root tissue, respectively. The new tissues produced are known collectively as the primary issues of the plant.

Axis: The second cervical vertebra which articulates with the atlas (the first cervical vertebra, which articulates with the skull). The articulation between the axis and atlas in reptiles, birds, and mammals permits side-to-side movement of the head. The body of the axis is elongated to form a peg, which extends into the ring of the atlas and acts as a pivot on which the atlas (and skull) can turn.

Azo compounds: A type of organic compound of the general formula RN:NR', where R and R' are aromatic groups. Azo compounds can be formed by coupling a diazonium compound with an aromatic phenol or amine. Most are coloured because of the -N:N-.

Backbone: Vertebral column. It is formed during development by the replacement of notochord. It is dorsal in position and encloses the spinal cord.

Back cross: Cross in which the hybrid is fertilized with one of its parents.

Bacteria: Unicellular plants which usually lack chlorophyll. Prokaryotic since they lack nuclear membrane and all other membrane bound structures. Reproduce by binary fission of great economic importance.

Bacteriods: Irregular, enlarged forms of rod-shaped bacteria, especially of the root-nodule forming species. These are ultimately absorbed by the cells of the root-nodule.

Bacteriology: Branch of biology dealing with the study of bacteria.

Bacteriophage: A virus which injects, parasitizes and kills bacteria; usually tadpole-shaped. Contains DNA or RNA. Has lytic or lysogenic cycle.

Bacteriorhiza: Symbiosis between a root and bacteria

Bacteriostatic: Substance which inhibits the growth and/or multiplication of bacteria, but does not kill them.

Bait: Foodstuff used for attracting pests. Usually mixed with poison to form a poison bait.

Balanced diet: Diet which contains adequate amounts of carbohydrates, fats, proteins, mineral salts, vitamins and water to meet the nutritional requirements of the body.

Balanced polymorphism: Presence of two or more distinct phenotypes in a population by natural selection It can occur as a result of disruptive selection or if the heterozygotes are more fit than either homozygote.

Balancers (halteres): Reduced hindwings of Diptera, shaped like short drumsticks e.g. house fly, mosquitoes, etc.

Balbiani rings (Puffs): Enlarged bands of giant chromosomes of the salivary glands. Such chromosomes are also called polytene chromosomes. Normally the chromosomes are invisible in the interphase but in the polytene chromosomes, first observed by Balbiani, the chromosomes are exceptionally thick.

Ballismus: Sudden, violent, purposeless movement of the distal part of' the arms or legs.

Barb: (Zool). The filaments, in a row on each side of the longitudinal axis, which together make the expanded part (vane) of a feather. (Bot.). A hooked or doubly hooked hair.

Barbule: (Bot). The inner row of peristome teeth in the capsule of some mosses (Zool). Minute filaments in a row at each side of barb of feathers. Those of one side bear hook, those of the other a groove. Barbules of adjacent barbs hook together and link barbs into a firm vane Downfeathers and ostrich feathers have no interlocking mechanism, so their barbs are free.

Bar eye: One of the mutations of *Drosophila* in which the number of optical units in the compound eve is greatly reduced.

Bark: Protective layer of cork((dead) cells, present on the outside of older stems and roots of woody plants, produced by activity of cork cambium. Bark may consist of cork only or when other layers of cork are formed at successively deeper levels, it may consist of alternating layers of cork and dead cortex or phloem tissue. Popularity regarded as consisting of everything outside the xylem.

Baroreceptor: A receptor sensitive to pressure. For example, the arterial baroreceptors respond to mean arterial pressure and pulse pressure.

Barr body: Inactivated X-chromosomes in the interphase nucleus also called sex-chromatin. One of the two X-chromosomes of a normal female becomes heterochromatic and appears as a chromatin body, stainable by orcein. The interphase nuclei of males do not have this body because each cell has only one X-chromosome. In cells with higher number of X-chromosomes, the number of Barr bodies increases correspondingly.

Bartholin's glands: Glands occurring in some mammals which lie on either side of the upper end of the vagina.

Basal body: (Blepharoplast; kinetosome). Centriole-like body found at the base of all cilia and flagella. It is essential for formation of cilia and flagella.

Basal metabolism: Minimum expenditure of energy for sustaining life, as during complete rest.

Basal ganglia: Several nuclei in the cerebral hemisphere that code and relay information associated with control of muscular movement.

Basal metabolism rate: Heat production per square metre of the body surface per hour at complete rest.

Basement membrane: Delicate intercellular membrane, which underlies most animal epithelia Besides the basal lamina, consists of mucopolysaccharide and very fine fibres.

Base ratio: In DNA the amount of adenine (A) and thymine (T) equals the amount of guanine (G) and cytosine (C). But the ratio of the amount of adenine plus thymine to that of guanine plus cytosine which is the base ratio, varies.

Basic dyes: Consist of a basic organic group (cation) which is the staining part, attached with an acid, usually inorganic. Used mainly for staining nucleic acids.

Basichromatin: Form of chromatin containing a fairly high proportion of nucleic acid and staining deeply with the basic dyes.

Basidia: Club-shaped cells present on the gills of mushrooms which produce haploid spores-basidio spores.

Basidiocarp: Fruiting body of the basidiomycete fungi (mushrooms etc.) excepting the rusts and smuts. The basidiocarp is the above ground portion of the fungus that is commonly referred to as the toadstool. The basidiocarp contains the basidial hyphae that, after meiosis, give rise to the basidiospores.

Basidiomycetes: A class of true fungi in which sexual spores, basidiospores, are borne externally on the end of specialized cells termed basidia. The basidia are formed on a fertile compact layer called the hymenium and each basidium usually contains four basidiospores.

Basidium: In basidiomycetes the cell in which fusion of haploid nuclei occurs during sexual reproduction, followed by meiosis and formation of four haploid basidiospores. Clubshaped or cylindrical cell, or divided into four cells; spores borne externally on minute stalks.

Basifixed: An anther which is attached by its base to the filament.

Basilar membrane: The membraneous portion in the inner car that separates the cochlear duct and scala tympani. It supports the organ of Corti.

Basipetal: Development of organs from the apex downwards so that the youngest structure are farthest from the apex. The term may also be applied to the movement of substances towards the base; for example the movement of auxin in shoot tissues.

Basophil: The polymor-phonuclear granulocytic leucocyte whose granules stain with basic dyes.

B-Cells: Lymphocyte which can become antibody-secreting plasma cells. They originate from bone marrow.

B.C.G: Bacillus Calmette Guerin vaccine was firstly introduced in France in 1908 by Calmette and Guerin. It produces significant immunity against the Tubercle bacillus.

Beak: The pointed protuberance or the snout (rostrum) of an animal such as beaked whale or the horn covered jaws of turtles and birds. The beak of birds is also called the bill.

Bee sting: The sting of a bee is a modified ovipositor on the hind end of the abdomen. It can be produced at will but normally rests in a pocket in the seventh abdominal segment. The piercing stylet consists of a number of 'valves' through which venom flows from the so-called acid gland.

Beeswax: Wax secreted by worker bees from the glands under abdomen. Used for making honeycomb. It is a mixture of esters and paraffin hydrocarbons, all of high molecular weight.

Beetles: Belong to order Coleoptera—the largest order of animal kingdom with 300,000 species. Forewings are horny and useless in flight. Mouth parts biting. The order also includes weevils and fireflies.

Bennettitales: Order of extinct Gymnospermae that flourished during the Mesozoic. Resembled cycads in external appearance and anatomy of stem and leaf.

Differed from them in possessing cones containing both micro and mega-sporophylls and in the form of these sporophylls.

Benzene (C6H6): A colourless liquid hydrocarbon with a characteristic odour. Benzene is highly toxic compound and continued inhalation of the vapour is harmful. It was originally isolated from coal tar and for many years this was the principal source of the compound. Contemporary manufacture is from hexane; petroleum vapour is passed over platinum at 500°C and at 10 atmosphere pressure:

$$C_6H_{14} ® C_6H_6 + 4H_2$$

Benzene is the simplest aromatic hydrocarbon. It shows characteristic substitution reactions despite the high degree of unsaturation implied in its formula.

Baciliariophyta: Diatoms, Unicellular or colonial algae, having ornamented cell wall and two halves, called valves, Plastids contain brown pigment isofucoxanthin as well as chlorophyll. The food reserves are fats and volutin. The cells are non-flagellate. Isogamous sexual reproduction. Some members form auxospores or endospores, and others produce-flagellated zooids which may be gametic in nature. Constitute part of the freshwater or marine plankton.

Backbone: Vertebral column. It is formed during development by the replacement of notochord. It is dorsal in position and encloses the spinal cord.

Back cross: Cross in which the hybrid is fertilized with one of its parents.

Bacteria: Unicellular plants which usually lack chlorophyll. Prokaryotic since they lack nuclear membrane and all other membrane bound structures. Reproduce by binary fission of great economic importance.

Bacteriods: Irregular, enlarged forms of rod-shaped bacteria, especially of the root-nodule forming species. These are ultimately absorbed by the cells of the root-nodule.

Bacteriology: Branch of biology dealing with the study of bacteria.

Bacteriophage: A virus which injects, parasitizes and kills bacteria; usually tadpole-shaped. Contains DNA or RNA. Has lytic or lysogenic cycle.

Bacteriorhiza: Symbiosis between a root and bacteria

Bacteriostatic: Substance which inhibits the growth and/or multiplication of bacteria, but does not kill them.

Bait: Foodstuff used for attracting pests. Usually mixed with poison to form a poison bait.

Balanced diet: Diet which contains adequate amounts of carbohydrates, fats, proteins, mineral salts, vitamins and water to meet the nutritional requirements of the body.

Balanced polymorphis: Presence of two or more distinct phenotypes in a population by natural selection It can occur as a result of disruptive selection or if the heterozygotes are more fit than either homozygote.

Balancers (halteres): Reduced hindwings of Diptera, shaped like short drumsticks e.g. house fly, mosquitoes, etc.

Balbiani rings (Puffs): Enlarged bands of giant chromosomes of the salivary glands. Such chromosomes are also called polytene chromosomes. Normally the chromosomes are invisible in the interphase but in the polytene chromosomes, first observed by Balbiani, the chromosomes are exceptionally thick.

Ballismus: Sudden, violent, purposeless movement of the distal part of' the arms or legs.

Barb (Zool): The filaments, in a row on each side of the longitudinal axis, which together make the expanded part (vane) of a feather. (Bot.). A hooked or doubly hooked hair.

Barbule (Bot): The inner row of peristome teeth in the capsule of some mosses (Zool). Minute filaments in a row at each side of barb of feathers. Those of one side bear hook, those of the other a groove. Barbules of adjacent barbs hook together and link barbs into a firm vane Downfeathers and ostrich feathers have no interlocking mechanism, so their barbs are free.

Bar eye: One of the mutations of *Drosophila* in which the number of optical units in the compound eve is greatly reduced.

Bark: Protective layer of cork((dead) cells, present on the outside of older stems and roots of woody plants, produced by activity of cork cambium. Bark may consist of cork only or when other layers of cork are formed at successively deeper levels, it may consist of alternating layers of cork and dead cortex or phloem tissue. Popularity regarded as consisting of everything outside the xylem.

Baroreceptor: A receptor sensitive to pressure. For example, the arterial baroreceptors respond to mean arterial pressure and pulse pressure.

Barr body: Inactivated X-chromosomes in the interphase nucleus also called sex-chromatin. One of the two X-chromosomes of a normal female becomes heterochromatic and appears as a chromatin body, stainable by orcein. The interphase nuclei of males do not have this body because each cell has only one X-chromosome. In cells with higher number of X-chromosomes, the number of Barr bodies increases correspondingly.

Bartholin's glands: Glands occurring in some mammals which lie on either side of the upper end of the vagina.

Basal body (Blepharoplast; kinetosome): Centriole-like body found at the base of all cilia and flagella. It is essential for formation of cilia and flagella.

Basal metabolism: Minimum expenditure of energy for sustaining life, as during complete rest.

Basal ganglia: Several nuclei in the cerebral hemisphere that code and relay information associated with control of muscular movement.

Basal metabolism rate: Heat production per square metre of the body surface per hour at complete rest.

Basement membrane: Delicate intercellular membrane, which underlies most animal epithelia Besides the basal lamina, consists of mucopolysaccharide and very fine fibres.

Base ratio: In DNA the amount of adenine (A) and thymine (T) equals the amount of guanine (G) and cytosine (C). But the ratio of the amount of adenine plus thymine to that of guanine plus cytosine which is the base ratio, varies.

Basic dyes: Consist of a basic organic group (cation) which is the staining part, attached with an acid, usually inorganic. Used mainly for staining nucleic acids.

Basichromatin: Form of chromatin containing a fairly high proportion of nucleic acid and staining deeply with the basic dyes.

Basidia: Club-shaped cells present on the gills of mushrooms which produce haploid spores-basidio spores.

Basidiocarp: Fruiting body of the basidiomycete fungi (mushrooms etc.) excepting the rusts and smuts. The basidiocarp is the above ground portion of the fungus that is commonly referred to as the toadstool. The basidiocarp contains the basidial hyphae that, after meiosis, give rise to the basidiospores.

Basidiomycetes: A class of true fungi in which sexual spores, basidio-spores, are borne externally on the end of specialized cells termed basidia. The basidia are formed on a fertile compact layer called the hymenium and each basidium usually contains four basidiospores.

Basidium: In basidiomycetes the cell in which fusion of haploid nuclei occurs during sexual reproduction, followed by meiosis and formation of four haploid basidiospores. Club-shaped or cylindrical cell, or divided into four cells; spores borne externally on minute stalks.

Basifixed: An anther which is attached by its base to the filament.

Basilar membrane: The membraneous portion in the inner car that separates the cochlear duct and scala tympani. It supports the organ of Corti.

Basipetal: Development of organs from the apex downwards so that the youngest structure are farthest from the apex. The term may also be applied to the movement of substances towards the base; for example the movement of auxin in shoot tissues.

Basophil: The polymor-phonuclear granulocytic leucocyte whose granules stain with basic dyes.

Bast fibre: Sclerenchymatous fibres in phloem. Bast. Used to mean 'phloem.'

B-Cells: Lymphocyte which can become antibody-secreting plasma cells. They originate from bone marrow.

B.C.G: Bacillus Calmette Guerin vaccine was firstly introduced in France in 1908 by Calmette and Guerin. It produces significant immunity against the Tubercle bacillus.

Beak: The pointed protuberance or the snout (rostrum) of an animal such as beaked whale or the horn covered jaws of turtles and birds. The beak of birds is also called the bill.

Bee sting: The sting of a bee is a modified ovipositor on the hind end of the abdomen. It can be produced at will but normally rests in a pocket in the seventh abdominal segment. The piercing stylet consists of a number of 'valves' through which venom flows from the so-called acid gland.

Beeswax: Wax secreted by worker bees from the glands under abdomen. Used for making honeycomb. It is a mixture of esters and paraffin hydrocarbons, all of high molecular weight.

Beetles: Belong to order Coleoptera—the largest order of animal kingdom with 300,000 species. Forewings are horny and useless in flight. Mouth parts biting. The order also includes weevils and fireflies.

Benign: Harmless, nonmalignant usually used for tumours of non-cancerous character.

Bennettitales: Order of extinct Gymnospermae that flourished during the Mesozoic. Resembled cycads in external appearance and anatomy of stem and leaf.

Benthos: Those animals and plants living on the bottom of sea or lake, from high water mark down to the deepest levels. The benthos is divided into littoral organisms (down to 200 metres deep) and deep water organisms.

Beriberi: Disease caused by the deficiency of thiamine (Vitamin B Symptoms are weakness, swelling and pain in legs, loss of appetite, enlarged heart, shortness of breath and paralysis.

Berry: A many-seeded fleshy fruit like tomato, brinjal, A few are one-seeded like date-palm, arecanut.

Beta-galactosidase: An enzyme which hydrolyzes the disaccharide, lactose.

Biceps: The muscle that causes flexion of the elbow.

Bicollateral bundle: A vascular bundle with phloem inside as well as outside the xylem on the same radius, e.g. in Cucurbitacae.

Bicuspid: Ending in two points as in bicuspid heart value of mammals. Two flaps of tissue guarding the opening between left and right auricle.

Biennial: A plant that completes its life cycle within two years. In the first year it produces foliage only and photosynthesizes. The food is stored during the winter in a swollen underground root or stem. In the second year, the food is used to produce flowers, fruits and seeds. Many important crops, such as carrot and turnip are biennials.

Bilaterally symmetrical: Organisms that can be halved in one and only one plane in such a way that the two halves are approximately mirror-images of each other. Usually this plane lies anteroposteriorly and dorsoventrally, thus separating similar right and left halves. Almost all freely moving animals are bilaterally symmetrical, e.g. all vertebrates, arthropods, and worm-shaped animals. Similar condition in flower called zygomorphy.

Bilateral cleavage: Cleavage that produces a bilaterally symmetrical arrangement of blastomeres lacking peculiar arrangement and the oblique divisions of spiral cleavage. It is found in Echinodermata, Chordata, etc.

Bile: Secretion of the liver of vertebrates, stored in the gall bladder and poured in the duodenum through bile duct. It is greenish yellow and alkaline to help neutralizing the acidity of the chyme. Chief constituents are proteins, water, inorganic salts, bile salts, lipids (cholesterol and lecithin) and bile pigments but no enzymes. Help in digestion of fats by.converting it into minute droplets (emulsification).

Bile duct: Duct from liver to duodenum invertebrates Transports bile juice.

Bile pigments: Pigments excreted in the bile as the products of break down of haemoglobin When haemoglobin is destroyed in the body, the protein portion, globin, is degraded to amino acids, while the porphyrin or haeme portion forms the bile pigments. It is easily reduced to the red-brown pigment bilirubin. Bilirubin is the major pigment in human bile, there being only slight traces of biliverdin, which is the chief pigment in the bile of birds.

Bile salts: Sodium carbonate, sodium glycocholate and sodium taurocholate present in the bile-juice. Sodium carbonate neutralizes the acidity of the chyme, the other two reduce the surface tension of water in the chyme and emulsify fats. Help in absorption of fats and fat-soluble vitamins (A, D, E, K).

Biliverdin: The green-coloured chief bile pigment in herbivores and birds. A precursor of bilirubin. Present in small amount in human bile.

Bill: The beak of a bird.

Binary fission Asexual method of reproduction in unicellular organisms involving the division of one cell into two.

Binocular vision Type of vision occurring in primates and many other vertebrates, in which eyeballs can be so directed that image of an object falls on both retinas. Extent to which eyes must be converged to bring the images on to a special part on each retina may be one of the mechanisms of distance judgement. Stereoscope vision (perception of the shape in depth) depends on two slightly, different images being received, because the two eyes look from different angles.

Binomial nomenclature The system of giving two names for each organism, first name a generic name and the second, a specific name. This system was first introduced by Linnaeus in 1758. Until then organisms were differently named in different parts of the world.

Bio-assay Quantitative estimation of biologically active substances by the amount of their actions in standardized conditions on living organisms or parts of organisms.

Biocatalyst A catalyst of organic origin. An enzyme which catalyses biochemical reactions.

Biochemistry: Chemistry of living organisms.

Biochemical Oxygen Demand (BOD) Standard measure for determining the level of organic pollution in a sample of water. It is the amount of oxygen used by microorganisms feeding on organic material over a given period of time. Sewage affluent must be diluted to campion with the statutory BOD before it can be disposed of into rivers.

Biocide A general poison or toxicant.

Biocoenosis: A community of plants (phytocoenosis) and animals (zoocoenosis).

Biodegradable: The substances that can be readily decomposed by the action of living organisms.

Biogenesis Principle of: The theory that a living thing can only originate from a living thing. It negates the theory of spontaneous generation of life. Foundation of this theory was laid by the work of Redi, Pasteur, etc.

Biogenetic law: The hypothesis that during the embryonic stage of development, an individual repeats all the evolutionary stages of its race. Briefly stated as embryogeny repeats phylogeny. Propounded by O.E. Haeckel.

Biogeography: Branch of biology that deals with the geographic distribution of plants and animals.

Biological clock: The internal mechanism of cells that regulates circadian rhythms and various other periodic cycles.

Biological control: Use of living organisms like natural predators and parasites to reduce pest population and control diseases.

Biology: The science of Eying beings. Two main branches are botany and zoology.

Bioluminiscence: Production of light by living organisms. It is found in many marine organisms, especially deepsea organisms, in some insects e.g. fire-life and certain bacteria. The light is produced as a result of a chemical reaction whereby the compound luciferin is oxidised. An enzyme, luciferase, catalyses the reaction in which ATP supplies the energy.

Biomagnification: Gradual concentration of pesticides, fungicides, etc., in the organisms higher in the food web.

Biomass: Total weight of all living organisms in an ecosystem.

Biomass: The major regional ecological community of plants and animals extending over large areas. For example, desert, tundra, tropical rain forest, etc.

Biometry: Application of mathematics, especially statistics for the study of having organisms.

Bionomics: (Zool) Study of the relationship of an organisms or population of organisms to environment.

Biont: An individual plant, independent, and capable of a separate existence.

Biophysics: Application of the knowledge of physics to the study of living things.

Biopoiesis: Origin of organisms from replicating molecules. DNA is the best example of a self-replicating molecule and is found in the chromosomes of all higher organisms.

Biopsy: A diagnostic procedure involving removal of a piece of tissue from a patient and its examination after sectioning and staining, for tumours, tuberculosis, cancer.

Bios factor: A substance essential for growth of a plant, especially yeast, obtained from the environment. It is a mixture of aneurin, biotin, and other substances.

Biosphere: That part of the earth and its atmosphere which is inhabited by living organisms.

Biosynthesis: Process of formation of a chemical substance by a living cell or an extract of a living cell.

Biosystematics: The area of systematics in which experimental taxonomic techniques are applied to investigate the relationship between taxa.

Biota: The plants and animals of a given region.

Biotic adaptation: Changes in the forms of physiology, presumed to have arisen as a result of competition with other organism.

Biotic climax: A climax community maintained by living organisms, e.g. grassland remaining static due to grazing of animals.

Biotic factors: The environmental influences due to activities of living organisms.

Biotic potential: An estimate of the maximum rate of increase of any animal species if left to itself and isolated from its natural enemies, diseases or other inhibiting factor. Normally this potential reproductive rate is very great (many millions per year in the case of insects), but in the struggle for existence a balance is maintained and the population of any species remains roughly constant.

Biotin: Water soluble vitamin of B-complex. Widely distributed in natural foods like egg yolk, kidney, liver, and yeast. Biotin is required as a coenzyme for carboxylation reaction in cellular metabolism.

Biotype: 1. Naturally occurring group of individuals with the same genetic composition, i.e. a clone or a pure line. 2. A physiologic race.

Bipinnaria larva: It is the larva of Asteroidea, the class which includes starfishes. The larva is free-swimming. Has ciliated bands, suggesting two rings.

Bipolar cells: The neurons in the retina that are postsynaptic to the rods and cones.

Biramous: Possessing two branches (forked), as biramous appendages of Crustacea.

Bisexual: Hermaphrodite or monoecious, where both the male and female sexes are borne by the same individual.

Bisporangiata: Strobilns or cone having both microsporophylls as well as megasporophylls with microsporangia and megasporangia, respectively.

Biuret reaction: Chemical reaction in which an alkaline solution of biuret gives a purple colour on the addition of cupric sulphate. Used as a biochemical test for protein and urea.

Bivalent A pair of homologous chromosomes when they pair up during meiosis. Pairing of homologous chromosomes is clearly seen during pachytene of meiosis I.

Bladder: (Bot) A modified (insectivorous) leaf, found on the stems of members of the bladderwort family that develops into a distended structure for trapping small invertebrates. (Zool.) A thinwalled muscular sac used as a temporary store for urine in most vertebrates (except birds and some reptiles). In mammals the urine enters the bladder directly fro m the ureters and is discharged to the urethra under the control of a sphincter muscle. It develops as an enlargement of the Wolffian duct or cloaca.

Bladder worm Cysticercus larva of tapeworm (*Taenia*) developed in the pig's flesh (measly pork). It is inactive in pig but develops into an adult tapeworm in man. It has a large vesicle and a scolex with sucker, hooks and rostellum.

Blastema: Mass of undifferentiated cells that later develops into an organ. One of the two main ways by which an animal regenerates a lost part is by initial formation of a blastema (e.g. limb or tail of newt, head of flatworm); the other way being by remodelling of remaining tissue (morphallaxis).

Blastocoel: The cavity of blastula formed at the end of cleavage during embryogenesis, providing space for gastrular movements to occur, and bounded by a single layer of blastomeres called blastoderm.

Blastocyst: Blastula like embryonic stage of placental mammals that implants in the wall of the uterus.

Blastoderm: Discoidal cap of cells above the blastocoel in a bird's egg. Its marginal area in which the cells remain undetached from the yolk and closed adherent to it is called the zone of junction.

Blastodisc: The protoplasm disc within which cleavage occurs in the eggs of fishes, reptiles and birds.

Blastogenesis: Transmission of inherited characters by the germplasm only.

Blastomere: One of the cells resulting due to cleavage of a fertilized egg of an animal.

Blastopore: Transitory opening on surface of embryo in gastrula stage, by which connects the archenteron with the exterior. In many animals the balastopore becomes the anus; in others it closes up at the end of gastrulation and the anus later breaks through at the same place or nearby. In some animals gastrulation movements carry 1 endoderm and mesoderm cells inwards at a particular place without forming an open pore (e.g. in a primitive streak); the place of inward migration may then be called virtual blastopore.

Blastostyle: A polyp from which medusae arise by budding in colonies of hydrozoan coelenterates like Obelia.

Blastotomy: Separation of cells or groups of cells of the blastula by any means.

Blastula: Early stage of embryonic development of animals in which a single layer of cells surrounds a fluid-filled cavity (blastocoel), thus forming a hollow ball.

Bleeder: An individual afflicted with haemophilia.

Blepharoplast A granule at the base of a flagellum or a cilium.

Blind spot: An area of the retina, where the optic nerve leaves the eye ball. It is insensitive to light a it is devoid of rods and cones. Therefore, no image is formed here.

Blood: A fluid connective tissue that circulates in vessels or spaces with endothelial lining. Usually contains respiratory pigments, serves as a medium of transport of oxygen, food materials, excretory products, hormones, etc. Cells suspended in fluid part (plasma). Present in Nemertea, Annelida, Arthropoda, Mollusca and Chordata. Three types of blood cells are red blood cells, white blood cells and platelets.

Blood Clotting: It is a mechanism for prevention of blood loss during an injury Blood oozing out of a wound soon forms a clot and further flow is prevented. Soluble blood protein fibrinogen is converted into fibrous fibrin by the action of enzyme thrombin. Thrombin is formed from blood protein, prothrombin by the action of thrombokinase. Thrombokinase is liberated by injured tissue or by platelets.

Blood corpuscles (cells): Three types of blood cells are red blood cells, white blood cells and platelets.

Blood groups: The classification of blood determined by the presence of antigens of either A, B, O or AB on the plasma membranes of RBCs and by the presence in the plasma of the anti-A or anti-B antibodies (or both or neither). Mixing bloods of different groups results in agglutination of the RBCs.

Blood plasma: The liquid that remains when all the cells are removed from the blood. It consists of 91% water and 7% proteins, which are albumins, globulins (mainly antibodies), prothrombin and fibrinogen. Plasma also contains ions of dissolved salts, especially sodium, potassium, chloride, bicarbonate; sulphate and phosphate,

Blood pressure: Usually refers to pressure of blood in the main arteries, which in normal human beings fluctuates roughly between 120 and 80

mm, of mercury, according to stage of heart beat (maximum at systole, minimum at diastole).

Blood sugar: 25% of the total glucose dissolved in the extracellular body fluids is the blood plasma in the form of D-glucose and it is equivalent to the glucose concentration in glomerular filtrate. Normal blood-sugar level is 50 to 80 mg per cent. A concentration above the normal maximum (120 mg per cent) is called hyperglycemia while its concentration below normal minimum is known, as hypoglycemia. Glucose in blood comes from digestive absorption, tissue fluids and secretion of glucose by the liver while it leaves the blood through diffusion.

Blood vessel Tube through which blood flows in the body.

Blubber: Thick layer of fat beneath the skin to prevent the loss of body heat into the surrounding cold water present in aquatic mammals like whales and seals as these have sparse or no hair on the body.

Blue sensitive cones: Cones containing photopigment most sensitive to blue light of 445 nm wavelength.

Body cavity: Internal space or a large fluid-filled cavity lying between the body wall and the internal organs; utilized as a temporary site for accumulation of excess fluids and waste, maturation of gametes, and providing space for enlargement of internal organs, Protozoans, sponges, Cnidarians, Acnidarians, flat-worms and proboscis worms lack a body cavity Pseudo-coelomates (Aschelminthes like rotifers and round worms) have a false body cavity which is a persistent blastocoel, not bounded by peritoneum. Other animals have a true coelom which arises as a cavity within the embryonic mesoderm.

Bolus: A lump of food that has been chewed and mixed with saliva.

Bone: Skeletal tissue of vertebrates. Consists of cells distributed in a matrix consisting largely of collagen fibres together with salts of calcium and phosphate. Bone salts 60% by weight of bone, is responsible for hardness; collagen for tensile strength. The cells are connected by fine channels which permeate the matrix. Larger channels contain blood vessels and nerves.

Bone marrow: A highly vascular, cellular substance in the central cavity of some bones. It is the site of synthesis of erythrocytes, some types of leucocytes and platelets.

Bordeaux mixture: A fungicide made of copper sulphate and quick lime. A common formula is 4 lb. copper sulphate, 4 lb. quick lime, dissolved in 50 gal. water.

Bordered pit: A thin area in the wall between two vessels or tracheids surrounded by overhanging rims of thickened secondary walls.

Botulism: A type of food poisoning caused by bacterium *Clostridium botulinum* which produces botulinus toxin. It blocks the release of acetyicholine from motor neuron terminals, thereby preventing the excitation of the muscle membrane. Early symptoms are vomitting, abdominal pain and difficulty of vision.

Bonquet stage: When chromosomes lie in loops with their ends near one part of the nuclear membrane during cell division.

Bowman's capsule: A blind sac at the beginning of the tubular component of a nephron. Its diameter in man is 1/5 to 1/10 mm.

Brachial: Pertaining to the fore limb (arm) or pectoral appendage.

Brachiopoda: Phylum of invertebrates including lampshells having a food-catching organ called lophophore Marine, have bivalved shell.

Brachydactyly: A condition in which the fingers or toes are abnormally short.

Brachypterous: Having short wings, that do not cover the abdomen.

Brachysclereids (stone cells) Short and isodiametric. Most common type of sclereids.

Bract: Small, leaflike structure at the base of a flower or an inflorescence.

Bracteol: A small leaf on the stalk of a flower. A small bract.

Brain: Anterior part of central nervous system originating from the embryonic ectoderm. Enclosed in a brain box in the vertebrates but otherwise present in almost all bilaterally symmetrical ani-mals. It controls and coordinates the actions of the body.

Brain stem: Vertebrate brain excluding forebrain and cerebellum

Branchial arch: Visceral arch lying between adjacent gill slits of fish i.e. third and following visceral arches in most fishes.

Branchial grooves: Series of external, paired grooves in the neck region of vertebrate embryo that correspond in position to the out-pocketings of the pharynx (gill pouches).

Branchiopoda: Subclass of small, freshwater crustaceans with flattened leaflike trunk appendages; with dense setae and gills on the coxa of the appendages, the abdomen is without appendage, e.g. fairy shrimps, tadpole shrimp, clamp shrimp, water flea.

Bronchiole: Small (less than 1 mm diameter) air conducting tube of tetrapod lung, arising as branch of a bronchus; terminating in alveoli. Smooth muscles in walls control size of lumen. Cartilage and mucus glands found in the bronchi are absent.

Bronchitis: Acute inflamation of the air passages.

Bronchus: Large air tube of tetrapod lung. Each lung has one large bronchus connecting it to trachea. Within the lung the bronchus branches into smaller and smaller bronchi and finally into bronchioles. Have cartilage plates, smooth muscles and mucus secreting glandcells in wall. Lining cells bear cilia beating towards mouth, which remove dust; etc.

Brush border: Closely packed regular sized microvilli on free surface of epithelium to form a brush border. It increases the surface area for absorption.

Bryophyta: Division of simple, green, non-vascular plants which are amphibious in nature. They are terrestrial, commonly found in moist habitats. Includes liverworts and mosses. The sex organs are multicellular and have a jacket layer of sterile cells. Gametophyte is the dominant generation. There is a distinct heteromorphic alternation of generation. The sporophyte is always attached to the gametophyte on which it is completely or partially dependent for nutrition.

Bryozoa, (Polyzoa): One of the smaller divisions of the animal kingdom; consists of moss, animals which are sessile and mostly colonial forming branching jellylike masses.

Bud (Bot): A compact undeveloped shoot consisting of a shortened stern and immature leaves or floral parts An outgrowth of an animal that is capable of asexual reproduction.

Buffer: Solution that resists change in p^H when acid or alkali is added to it. Blood is a buffer.

Bug: Insect members of the order Hemiptera, a large order including both winged and wingless species. The most characteristic feature is the very long proboscis adapted for piercing and sucking.

Bulb: A condensed, underground stem bearing fleshy leaves It is responsible for food storage and vegetative reproduction.

Bulbil: A small bulb-like organ of vegetative reproduction that may form in the leaf axil, an inflorescence, or at a stem base. Modified axillary vegetative or floral bud.

Bulbo-urethral glands: Paired glands in mammals whose ducts open into urethra near the base of penis.

Bundle sheath: Layer(s) of cells that surround a vascular bundle. May be parenchymatous or sclerenchymatous.

Bursa of Fabricius: The structure formed from the cloaca of birds. It is the source of lymphocytes.

Beriberi: Disease caused by the deficiency of thiamine (Vitamin B Symptoms are weakness, swelling and pain in legs, loss of appetite, enlarged heart, shortness of breath and paralysis.

Berry: A many-seeded fleshy fruit like tomato, brinjal, A few are one-seeded like date-palm, arecanut.

Beta-galactosidase: An enzyme which hydrolyzes the disaccharide, lactose.

Biceps: The muscle that causes flexion of the elbow.

Bicollateral bundle: A vascular bundle with phloem inside as well as outside the xylem on the same radius, e.g. in Cucurbitacae.

Bicuspid: Ending in two points as in bicuspid heart value of mammals. Two flaps of tissue guarding the opening between left and right auricle.

Biennial: A plant that completes its life cycle within two years. In the first year it produces foliage only and photosynthesizes. The food is stored during the winter in a swollen underground root or stem. In the second year, the food is used to produce flowers, fruits and seeds. Many important crops, such as carrot and turnip are biennials.

Bilaterally symmetrical: Organisms that can be halved in one and only one plane in such a way that the two halves are approximately mirror-images of each other. Usually this plane lies anteroposteriorly and dorsoventrally, thus separating similar right and left halves. Almost all freely moving animals are bilaterally symmetrical, e.g. all vertebrates, arthropods, and worm-shaped animals. Similar condition in flower called zygomorphy.

Bilateral cleavage: Cleavage that produces a bilaterally symmetrical arrangement of blastomeres lacking peculiar arrangement and the oblique divisions of spiral cleavage. It is found in Echinodermata, Chordata, etc.

Bile: Secretion of the liver of vertebrates, stored in the gall bladder and poured in the duodenum through bile duct. It is greenish yellow and alkaline to help neutralizing the acidity of the chyme. Chief constituents are proteins, water, inorganic salts, bile salts, lipids (cholesterol and lecithin) and bile pigments but no enzymes. Help in digestion of fats by converting it into minute droplets (emulsification).

Bile duct: Duct from liver to duodenum invertebrates Transports bile juice.

Bile pigments: Pigments excreted in the bile as the products of break down of haemoglobin When haemoglobin is destroyed in the body, the protein portion, globin, is degraded to amino acids, while the porphyrin or haeme portion forms the bile pigments. It is easily reduced to the red-brown pigment bilirubin. Bilirubin is the major pigment in human bile, there being only slight traces of biliverdin, which is the chief pigment in the bile of birds.

Bile salts: Sodium carbonate, sodium glycocholate and sodium taurocholate present in the bile-juice. Sodium carbonate neutralizes the acidity of the chyme, the other two reduce the surface tension of water in the chyme and emulsify fats. Help in absorption of fats and fat-soluble vitamins (A, D, E, K).

Bilirubin: A red pigment found in bile.

Biliverdin: The green-coloured chief bile pigment in herbivores and birds. A precursor of bilirubin. Present in small amount in human bile.

Bill: The beak of a bird.

Binary fission: Asexual method of reproduction in unicellular organisms involving the division of one cell into two.

Binocular vision: Type of vision occurring in primates and many other vertebrates, in which eyeballs can be so directed that image of an object falls on both retinas. Extent to which eyes must be converged to bring the images on to a special part on each retina may be one of the mechanisms of distance judgement. Stereoscope vision (perception of the shape in depth) depends on two slightly, different images being received, because the two eyes look from different angles.

Binomial nomenclature: The system of giving two names for each organism, first name a generic name and the second, a specific name. This system was first introduced by Linnaeus in 1758. Until then organisms were differently named in different parts of the world.

Bio-assay: Quantitative estimation of biologically active substances by the amount of their actions in standardized conditions on living organisms or parts of organisms.

Biocatalyst: A catalyst of organic origin. An enzyme which catalyses biochemical reactions.

Biochemistry: Chemistry of living organisms.

Biochemical Oxygen Demand (BOD): Standard measure for determining the level of organic pollution in a sample of water. It is the amount of oxygen used by microorganisms feeding on organic material over a given period of time. Sewage affluent must be diluted to campion with the statutory BOD before it can be disposed of into rivers.

Biocide: A general poison or toxicant.

Biocoenosis: A community of plants (phytocoenosis) and animals (zoocoenosis).

Biodegradable: The substances that can be readily decomposed by the action of living organisms.

Biogenesis Principle of: The theory that a living thing can only originate from a living thing. It negates the theory of spontaneous generation of life. Foundation of this theory was laid by the work of Redi, Pasteur, etc.

Biogenetic law: The hypothesis that during the embryonic stage of development, an individual repeats all the evolutionary stages of its race. Briefly stated as embryogeny repeats phylogeny. Propounded by O.E. Haeckel.

Biogeography: Branch of biology that deals with the geographic distribution of plants and animals.

Biological clock: The internal mechanism of cells that regulates circadian rhythms and various other periodic cycles.

Biological control: Use of living organisms like natural predators and parasites to reduce pest population and control diseases.

Biology: The science of Eying beings. Two main branches are botany and zoology.

Bioluminiscence: Production of light by living organisms. It is found in many marine organisms, especially deepsea organisms, in some insects: e.g. fire-life and certain bacteria. The light is produced as a result of a chemical reaction whereby the compound luciferin is oxidised. An enzyme, luciferase, catalyses the reaction in which ATP supplies the energy.

Biomagnification: Gradual concentration of pesticides, fungicides, etc., in the organisms higher in the food web.

Biomass: Total weight of all living organisms in an ecosystem.

Biomass: The major regional ecological community of plants and animals extending over large areas. For example, desert, tundra, tropical rain forest, etc.

Biometry: Application of mathematics, especially statistics for the study of having organisms.

Bionomics: (Zool) Study of the relationship of an organisms or population of organisms to environment.

Biont: An individual plant, independent, and capable of a separate existence.

Biophysics: Application of the knowledge of physics to the study of living things.

Biopoiesis: Origin of organisms from replicating molecules. DNA is the best example of a self-replicating molecule and is found in the chromosomes of all higher organisms.

Biopsy: A diagnostic procedure involving removal of a piece of tissue from a patient and its examination after sectioning and staining, for tumours, tuberculosis, cancer.

Bios factor: A substance essential for growth of a plant, especially yeast, obtained from the environment. It is a mixture of aneurin, biotin, and other substances.

Biosphere: That part of the earth and its atmosphere which is inhabited by living organisms.

Biosynthesis: Process of formation of a chemical substance by a living cell or an extract of a living cell.

Biosystematics: The area of systematics in which experimental taxonomic techniques are applied to investigate the relationship between taxa.

Biota: The plants and animals of a given region.

Biotic adaptation: Changes in the forms of physiology, presumed to have arisen as a result of competition with other organism.

Biotic climax: A climax community maintained by living organisms, e.g. grassland remaining static due to grazing of animals.

Biotic factors: The environmental influences due to activities of living organisms.

Biotic potential: An estimate of the maximum rate of increase of any animal species if left to itself and isolated from its natural enemies, diseases or other inhibiting factor. Normally this potential reproductive rate is very great (many millions per year in the case of insects), but in the struggle for existence a balance is maintained and the population of any species remains roughly constant.

Biotin: Water soluble vitamin of B-complex. Widely distributed in natural foods like egg yolk, kidney, liver, and yeast. Biotin is required as a coenzyme for carboxylation reaction in cellular metabolism.

Biotype: 1. Naturally occurring group of individuals with the same genetic composition, i.e. a clone or a pure line. 2. A physiologic race.

Bipinnaria larva: It is the larva of Asteroidea, the class which includes starfishes. The larva is free-swimming. Has ciliated bands, suggesting two rings.

Bipolar cells: The neurons in the retina that are postsynaptic to the rods and cones.

Biramous: Possessing two branches (forked), as biramous appendages of Crustacea.

Bisexual: Hermaphrodite or monoecious, where both the male and female sexes are borne by the same individual.

Bisporangiata: Strobilns or cone having both microsporophylls as well as megasporophylls with microsporangia and megasporangia, respectively.

Biuret reaction: Chemical reaction in which an alkaline solution of biuret gives a purple colour on the addition of cupric sulphate. Used as a biochemical test for protein and urea.

Bivalent: A pair of homologous chromosomes when they pair up during meiosis. Pairing of homologous chromosomes is clearly seen during pachytene of meiosis I.

Bladder: (Bot) A modified (insectivorous) leaf, found on the stems of members of the bladderwort family that develops into a distended structure for trapping small invertebrates. (Zool.) A thinwalled muscular sac used as a temporary store for urine in most vertebrates (except birds and some reptiles). In mammals the urine enters the bladder directly fro m the ureters and is discharged to the urethra under the control of a sphincter muscle. It develops as an enlargement of the Wolffian duct or cloaca.

Bladder worm: Cysticercus larva of tapeworm (*Taenia*) developed in the pig's flesh (measly pork). It is inactive in pig but develops into an adult tapeworm in man. It has a large vesicle and a scolex with sucker, hooks and rostellum.

Blastema: Mass of undifferentiated cells that later develops into an organ. One of the two main ways by which an animal regenerates a lost part is by initial formation of a blastema (e.g. limb or tail of newt, head of flatworm); the other way being by remodelling of remaining tissue.

Blastocoel: The cavity of blastula formed at the end of cleavage during embryogenesis, providing space for gastrular movements to occur, and bounded by a single layer of blastomeres called blastoderm.

Blastocyst: Blastula like embryonic stage of placental mammals that implants in the wall of the uterus.

Blastoderm: Discoidal cap of cells above the blastocoel in a bird's egg. Its marginal area in which the cells remain undetached from the yolk and closed adherent to it is called the zone of junction.

Blastodisc: The protoplasm disc within which cleavage occurs in the eggs of fishes, reptiles and birds.

Blastogenesis: Transmission of inherited characters by the germplasm only.

Blastomere: One of the cells resulting due to cleavage of a fertilized egg of an animal.

Blastopore: Transitory opening on surface of embryo in gastrula stage, by which connects the archenteron with the exterior. In many animals the balastopore becomes the anus; in others it closes up at the end of gastrulation and the anus later breaks through at the same place or nearby. In some animals gastrulation movements carry 1 endoderm and mesoderm cells inwards at a particular place without forming an open pore (e.g. in a primitive streak); the place of inward migration may then be called virtual blastopore.

Blastostyle: A polyp from which medusae arise by budding in colonies of hydrozoan coelenterates like Obelia.

Blastotomy: Separation of cells or groups of cells of the blastula by any means.

Blastula: Early stage of embryonic development of animals in which a single layer of cells surrounds a fluid-filled cavity (blastocoel), thus forming a hollow ball.

Bleeder: An individual afflicted with haemophilia.

Blepharoplast: A granule at the base of a flagellum or a cilium.

Blind spot: An area of the retina, where the optic nerve leaves the eye ball. It is insensitive to light a it is devoid of rods and cones. Therefore, no image is formed here.

Blood: A fluid connective tissue that circulates in vessels or spaces with endothelial lining. Usually contains respiratory pigments, serves as a medium of transport of oxygen, food materials, excretory products, hormones, etc. Cells suspended in fluid part (plasma). Present in Nemertea, Annelida, Arthropoda, Mollusca and Chordata. Three types of blood cells are red blood cells, white blood cells and platelets.

Blood Clotting: It is a mechanism for prevention of blood loss during an injury Blood oozing out of a wound soon forms a clot and further flow is prevented. Soluble blood protein fibrinogen is converted into fibrous fibrin by the action of enzyme thrombin. Thrombin is formed from blood protein, prothrombin by the action of thrombokinase. Thrombokinase is liberated by injured tissue or by platelets.

Blood corpuscles (cells): Three types of blood cells are red blood cells, white blood cells and platelets.

Blood groups: The classification of blood determined by the presence of antigens of either A, B, O or AB on the plasma membranes of RBCs and by the presence in the plasma of the anti-A or anti-B antibodies (or both or neither). Mixing bloods of different groups results in agglutination of the RBCs.

Blood plasma: The liquid that remains when all the cells are removed from the blood. It consists of 91% water and 7% proteins, which are albumins, globulins (mainly antibodies), prothrombin and fibrinogen. Plasma also contains ions of dissolved salts, especially sodium, potassium, chloride, bicarbonate; sulphate and phosphate.

Blood pressure: Usually refers to pressure of blood in the main arteries, which in normal human beings fluctuates roughly between 120 and 80 mm, of mercury, according to stage of heart beat (maximum at systole, minimum at diastole).

Blood sugar: 25% of the total glucose dissolved in the extracellular body fluids is the blood plasma in the form of D-glucose and it is equivalent to the glucose concentration in glomerular filtrate. Normal blood-sugar level is 50 to 80 mg per cent. A concentration above the normal maximum (120 mg per cent) is called hyperglycemia while its concentration below normal minimum is known, as hypoglycemia. Glucose in blood comes from digestive absorption, tissue fluids and secretion of glucose by the liver while it leaves the blood through diffusion.

Blood vessel: Tube through which blood flows in the body.

Blubber: Thick layer of fat beneath the skin to prevent the loss of body heat into the surrounding cold water present in aquatic mammals like whales and seals as these have sparse or no hair on the body.

Blue sensitive cones: Cones containing photopigment most sensitive to blue light of 445 nm wavelength.

Body cavity: Internal space or a large fluid-filled cavity lying between the body wall and the internal organs; utilized as a temporary site for accumulation of excess fluids and waste, maturation of gametes, and providing space for enlargement of internal organs, Protozoans, sponges, Cnidarians, Acnidarians, flat-worms and proboscis worms lack a body cavity Pseudocoelomates (Aschelminthes like rotifers and round worms) have a false body cavity which is a persistent blastocoel, not bounded by peritoneum. Other animals have a true coelom which arises as a cavity within the embryonic mesoderm.

Bolus: A lump of food that has been chewed and mixed with saliva.

Bone: Skeletal tissue of vertebrates. Consists of cells distributed in a matrix consisting largely of collagen fibres together with salts of calcium and phosphate. Bone salts 60% by weight of bone, is responsible for hardness; collagen for tensile strength. The cells are connected by fine channels which permeate the matrix. Larger channels contain blood vessels and nerves.

Bone marrow: A highly vascular, cellular substance in the central cavity of some bones. It is the site of synthesis of erythrocytes, some types of leucocytes and platelets.

Bordeaux mixture: A fungicide made of copper sulphate and quick lime. A common formula is 4 lb. copper sulphate, 4 lb. quick lime, dissolved in 50 gal. water.

Bordered pit: A thin area in the wall between two vessels or tracheids surrounded by overhanging rims of thickened secondary walls.

Boreal forest: Northern spruce-fir pine forest; the taiga.

Botulism: A type of food poisoning caused by bacterium *Clostridium botulinum* which produces botulinus toxin. It blocks the release of acetyicholine from motor neuron terminals, thereby preventing the excitation of the muscle membrane. Early symptoms are vomitting, abdominal pain and difficulty of vision.

Bonquet stage: When chromosomes lie in loops with their ends near one part of the nuclear membrane during cell division.

Bowman's capsule: A blind sac at the beginning of the tubular component of a nephron. Its diameter in man is 1/5 to 1/10 mm.

Brachial: Pertaining to the fore limb (arm) or pectoral appendage.

Brachiopoda: Phylum of invertebrates including lampshells having a food-catching organ called lophophore Marine, have bivalved shell.

Brachydactyly: A condition in which the fingers or toes are abnormally short.

Brachypterous: Having short wings, that do not cover the abdomen.

Brachysclereids (stone cells): Short and isodiametric. Most common type of sclereids.

Bract: Small, leaflike structure at the base of a flower or an inflorescence.

Bracteole: A small leaf on the stalk of a flower. A small bract.

Brain: Anterior part of central nervous system originating from the embryonic ectoderm. Enclosed in a brain box in the vertebrates but otherwise present in almost all bilaterally symmetrical animals. It controls and coordinates the actions of the body.

Brain stem: Vertebrate brain excluding forebrain and cerebellum

Branchial arch: Visceral arch lying between adjacent gill slits of fish i.e third and following visceral arches in most fishes.

Branchial grooves: Series of external, paired grooves in the neck region of vertebrate embryo that correspond in position to the out-pocketings of the pharynx (gill pouches).

Branchiopoda: Subclass of small, freshwater crustaceans with flattened leaflike trunk appendages; with dense setae and gills on the coxa of the appendages, the abdomen is without appendage. e.g. fairy shrimps, tadpole shrimp, clamp shrimp, water flea.

Bronchiole: Small (less than 1 mm diameter) air conducting tube of tetrapod lung, arising as branch of a bronchus; terminating in alveoli. Smooth muscles in walls control size of lumen. Cartilage and mucus glands found in the bronchi are absent.

Bronchus: Large air tube of tetrapod lung. Each lung has one large bronchus connecting it to trachea. Within the lung the bronchus branches into smaller and smaller bronchi and finally into bronchioles. Have cartilage plates, smooth muscles and mucus secreting glandcells in wall. Lining cells bear cilia beating towards mouth, which remove dust; etc.

Brush border: Closely packed regular sized microvilli on free surface of epithelium to form a brush border. It increases the surface area for absorption.

Bryophyta: Division of simple, green, non-vascular plants which are amphibious in nature. They are terrestrial, commonly found in moist habitats. Includes liverworts and mosses. The sex organs are multicellular and have a jacket layer of sterile cells. Gametophyte is the dominant generation. There is a distinct heteromorphic alternation of generation. The sporophyte is always attached to the gametophyte on which it is completely or partially dependent for nutrition.

Bryozoa, (Polyzoa): One of the smaller divisions of the animal kingdom; consists of moss, animals which are sessile and mostly colonial forming branching jellylike masses.

Bud (Bot): A compact undeveloped shoot consisting of a shortened stern and immature leaves or floral parts An outgrowth of an animal that is capable of asexual reproduction.

Buffer: Solution that resists change in pH when acid or alkali is added to it. Blood is a buffer.

Bug: Insect members of the order Hemiptera, a large order including both winged and wingless species. The most characteristic feature is the very long proboscis adapted for piercing and sucking.

Bulb: A condensed, underground stem bearing fleshy leaves It is responsible for food storage and vegetative reproduction.

Bulbil: A small bulb-like organ of vegetative reproduction that may form in the leaf axil, an inflorescence, or at a stem base. Modified axillary vegetative or floral bud.

Bulbo-urethral glands: Paired glands in mammals whose ducts open into urethra near the base of penis.

Bundle sheath: Layer(s) of cells that surround a vascular bundle. May be parenchymatous or sclerenchymatous.

C

Caenozoic (Cenozoic): Present geological era, beginning some 65 million years ago, and divided into two periods, the Tertiary and the Quaternary. It is characterised by the rise of mammals and flowering plants.

Caffeine: White crystalline purine: Occurs in tea-leaves, coffee beans, and other plant materials. Has a powerful action on the heart.

Calcarea: A class of phylum Porifera, including marine sponges, have calcareous spicules, e.g. *Leucosolenia* and *Sycon*. Canal system may be ascon type or sycon type.

Calcareous: Composed of or containing calcium carbonate.

Calcicole: Plants that thrive on neutral to alkaline chalk, carboniferous or limestone soils, such as marls.

Calciferol: Vitamin D. Formed by the action of ultraviolet radiation on ergosterol. Controls the deposition of calcium compounds in the body; deficiency causes rickets.

Calcifuse: Plants that grow best on acid soils.

Calcitonin (thyrocalcitonin): Vertebrate hormone secreted by parafollicular cells of mammalian thyroid, or from separate gland in other vertebrates. Controls blood level of calcium by opposing the action of the parathyroid.

Calf: 1. The young of cattle and related wild species. 2. The flesh mass formed by the muscles at the back of the leg.

Callose: A carbohydrate, which is deposited seasonally or permanently on sieve plates, leading them to stop functioning. It is also found in the microspore mother cell walls and in the cells of some algae. It is digested by the enzyme callase.

Callus: (Bot) A mass of undifferentiated parenchyma cells formed in response to wounding. In tissue cultures, callus can be induced by various hormone treatments. Adventitious shoots and roots often differentiate from calluses, a phenomenon exploited in the rooting of cuttings, (Zool.) 1. Thickened horny mass found in the outer layer of the skin formed as a result of' continued pressure of friction. 2. Tissue formed around the fragments of a broken bone which develops into a bone to heal the fracture.

Calorie: Amount of heat necessary to raise the temperature of 1 gm of water by 1°C (from say 14.5°C to 15.5°C).

Calyptra: A layer of cells derived from venter of tow archegonium. It covers the developing sporophyte. In bryophytes it ruptures as the seta elongates, being taken up as a hood over the capsule in mosses. The presence of the calyptra is necessary for the proper development of the capsule is mosses and the embryo in ferns.

Calyptrogen: A layer of meristematic cells over the root apex in some plants (e.g. grasses) that gives rise to the root cap.

Calyx: Outemost floral whorl; consists of green leaflike sepals which enclose flower during bud stage.

Cambium: Lateral meristem. Actively dividing cells between xylem and phloem. 11 forms additional layers of xylem and phloem, This phenomenon is known as secondary growth and occurs only in dicots.

Cambrian: Geological period which lasted from 570 till 500 million years ago.

Campylotropous: The ovule is curved in the form of the horseshoe. The microphyle and chalaza do not lie in a straight line.

Canada balsam: Gum commonly used for making permanent microscopical preparations. The object is placed in a thin layer of balsam solution between coverslip and slide. The balsam dries hard, and because its refractive index is like that of proteins and other constituents of biological objects, it makes them transparent.

Cancer: It is uncontrolled mitosis. Cancer cells continue to grow and divide, and form a mass known as malignant tumour. It invades the surrounding tissue and disrupts the structure and function of organs, leading to the death of the organism.

Canine tooth: Dog tooth of mammals, well developed in carnivores, absent in many herbivores like rabbit, hare, rat, squirrel and ungulates and reduced in some others; present on either side of each jaw in between the incisors and premolars, some times enlarged to tusks as in wild boar.

Cantheridin: A drug extracted from the blister beetle or spanish fly, *Lytta vesicatoria,* and others of the family Meloidae.

Capillary (blood): A very fine tubular branch with unicellular wall which connects a venule with arteriole. Through this diffusion takes place.

Capillitium: (1) Tubular protoplasmic threads in fruit bodies of Myxomycetes (slime moulds). Assist in discharge of spores in some species by their movements in response to changes in humidity. (2) Sterile hyphae in fruit bodies of certain fungi, e.g. puff-balls.

Capitulum: Racemose inflorescence with compressed disc-like pedicel bearing sessile flowers oldest at the margins and younger in the centre. Found in Compositae family.

Capsid: The protein coat of a virus, surrounding the nucleic acid. A capsid is present only in the inert extracellular stage of the life cycle, Capsids are composed of subunits called capsomeres. Involved in penetration of host cell.

Capsomere: The protein molecules which surround the nucleic acid., central core of a virus.

Capsule: (Bot) (1) in flowering plants, dry dehiscent fruit developed compound ovary; opening to liberate seeds in various ways. (2) In liverworts and mosses, organ, within which spores are formed. Part of sporophyte, (3) In some kinds of bacteria, gelatinous envelope surrounding the cell wall. (Zool). Connective tissue investment of an organ, providing mechanical support.

Captaculum: A tentacle-like outgrowth from head of tooth-shell molluscs, for food capturing.

Carapace: (1) Shield of exoskeleton covering part of the body of some Arthropoda, e.g. crabs (2) Dorsal part 'shell' of Chelonia, consisting of plates beneath the skin fused with ribs of vertebral column.

Carbohydrate: An organic compound composed of a chain of carbon atoms to which hydrogen and oxygen are attached in the ratio 2:1. Its general formula is $C_x (H_2O)y$. Carbohydrates include sugars, starches, cellulose and glycogen.

Carbon cycle: Circulation of carbon between living organisms and the environment. Carbon dioxide in the atmosphere is taken up by autotrophic organism (mainly green plants) and incorporated into carbohydrates. The Carbohydrates are the food source of the heterotrophs. All organisms return carbon dioxide to the air as a product of respiration and of decay. The burning of fossil fuels also releases CO_2.

Carboniferous: Geological period, lasted approximately from 345 till 280 million years ago.

Carboxyhaemoglobin: Compound formed by the union of haemoglobin and carbon monoxide.

Carboxylic acid: An organic compound of general formula RCOOH, where R is an organic group and -COOH is the carboxylate group. Many carboxylic acids are of biochemical importance.

Cardinal veins: Two parts of veins found in fish and tetrapod embryos that carry deoxygenated blood back to the heart. 'The anterior cardinal veins serve the head while the posterior cardinal veins serve the rest of the body. They unite to form the common cardinal vein (Cuvierian duct).

Cardiovascular: Pertaining to the heart and blood vessels.

Cardo: Hinge-like part of an organ but particularly the basal segment of an insect's maxilla by which it articulates with the head.

Carina: Keel of birds. Well developed in modern flying birds but reduced in the flightless birds like ostrich.

Carnassial tooth: Molar or premolar, modified for tearing flesh. In many Carnivora (e.g. dog, cat), lower first molar and upper last premolar.

Carnivora: Order of placental mammals comprising flesh-eating forms (cats, wolves, bears, seals). Most members of the order (but not seals) have well developed incisor and canine teeth and usually a pair of carnassial teeth on each side. Possess claws, often retractile.

Carotenoids: Group of yellow orange and red plant pigments located in chloroplasts and in plastids in other plant parts where chlorophyll is absent, e.g. carrot, many flowers and in photosynthetic lamellae of blue-green algae and some bacteria. Also occur in some fungi. Increases in concentration in many ripening fruits, e.g. tomato. Assist in photosynthesis by absorbing light and passing energy on to chlorophyll.

Carotid artery: One of the pair of blood vessels that supplies oxygenated blood to the head and neck. Derived from the third aortic arch, which in tetrapods forms the carotid arch that arises from the aorta as the common carotid artery. Branches in the neck region into an internal and external carotid artery.

Carotid body: A vascular structure at the base of the external carotid artery that contains chemoreceptors. These monitor carbon dioxide and oxygen concentrations and pH of the blood.

Carpal bones: The small bones of the wrist in man and the similar group in the forelegs of animals. Eight bones in man.

Carpet: Female reproductive organ of flowering plants. Consists of ovary, stigma and style.

Carpospore: Uninucleate spore formed from the direct or indirect division of the zygote of the red algae,

Carpophore: (1) Raised part of' the receptacle bearing carpels and stamens: (2) Stalk of the sporocarp (3) Forked stalk from which the mericarps of some umbelliferae are suspended.

Carpus: The region of forelimb of tetrapod vertebrates containing carpel bones, roughly the wrist.

Carrier (Bot.): A transport protein within the membrane that binds temporarily a with a molecule to facilitate the diffusion of the: latter across the membrane. Also refers to an individual heterozygous for a recessive gene in which the character is not phenotypically. (Zool.). A person who carries the specific organisms of any disease but without showing any sign of disease transmits disease through contact with others.

Cartilage: Skeletal tissue of vertebrates consisting of rounded cells scattered in a matrix, with numerous collagen fibres. Devoid of blood vessels in adult.

Cartilage-bone: (replacing bone.) A bone which replaces embryonic cartilage, e.g. limb bones; hip girdle; vertebral column; parts of skull such as auditory capsule.

Caruncle: A fleshy whitish outgrowth formed at the micropyle and hilum of some of the seeds, e.g. castor, bean, etc. It is derived from the outer integument.

Caryopsis: An achenial (simple, dry and indehiscent) one-seeded fruit in which the seed coat and the fruit coat are fused, as in Gramineae.

Casparian thickenings: Thickening due to the deposition of lignin or suberin on the walls of endodermis. These are impervious to water.

Castes: Morphologically and functionally distinct types within a colony of social insects. Particularly well developed in Hymenoptera (bees, ants and wasps) and in Isoptera (termites). Among bees the three chief castes are the queen (fertile female), workers (sterile females) and drones (fertile males).

Catabolism: (Katabolism). Metabolic reactions involved in the break-down of complex molecules to simple compounds. The function of catabolic reactions is to provide energy, which is used in the synthesis of new structures, for work, for transmission of nerve impulses, and for the maintenance of functional efficiency.

Catalase: Enzyme that breaks down toxic hydrogen peroxide to water and oxygen. During this reaction it is slowly destroyed itself.

Catalyst: Substance that accelerates the rate of chemical reaction without being used up in the process.

Cataract: Any opacity which develops in the crystalline lens of the eye. This is mostly due to the yellowing of the lens with age.

Caterpillar: A polyped or eruciform larva; a soft bodied larva having, in addition to the six true legs on the thorax, a number of prolegs or false legs on the abdomen. Such larvae occur in butterflies and moths, Caterpillars are herbivorous with powerful side-ways-moving jaws.

Cauliflory: Production of flowering shoots on older, thickened, leafless branches or main trunk. Common among the angiosperm trees of lower stories of canopy in tropical forests, e.g. cocoa.

C-banding: A method for staining chromosomes differently showing locations of heterochromatin in the stained banded regions of the chromosomes.

Cell: The structural and functional unit of living organisms. Cell consists of cell wall, protoplasm, nucleus and a large Vacuole in plants.

In animal cells, cell wall is absent and is bounded on the outer side by the plasma membrane only. The vacuoles are practically absent in animal cell or are small in size, if present. The protoplasm is lined with plasma membrane on the outer side. The cytoplasm has several metabolically active and inactive inclusions within. There are active inclusions, called organelles, e.g. mitochondria, chloroplast (in green cells of plants only), ribosomes, etc. The inactive inclusions are the food-reserves, etc. The cytoplasm consists of a network or a canal system called endoplasmic reticulum.

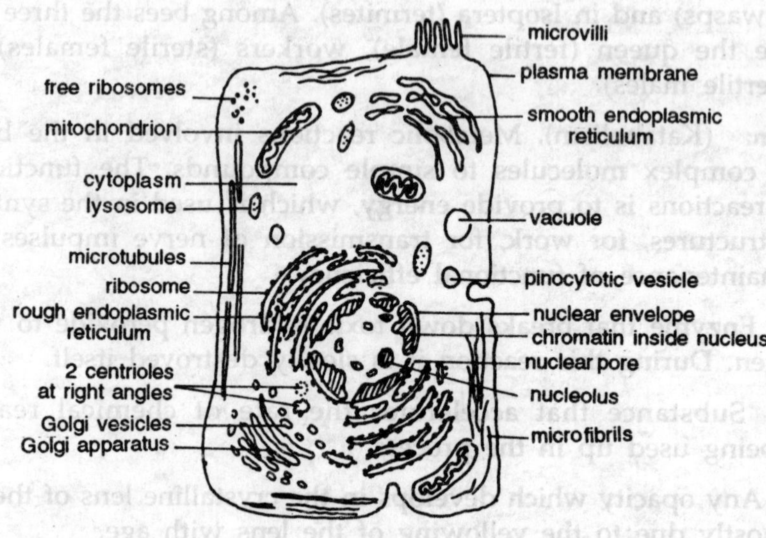

Generalised animal cell as seen under the electron microscope

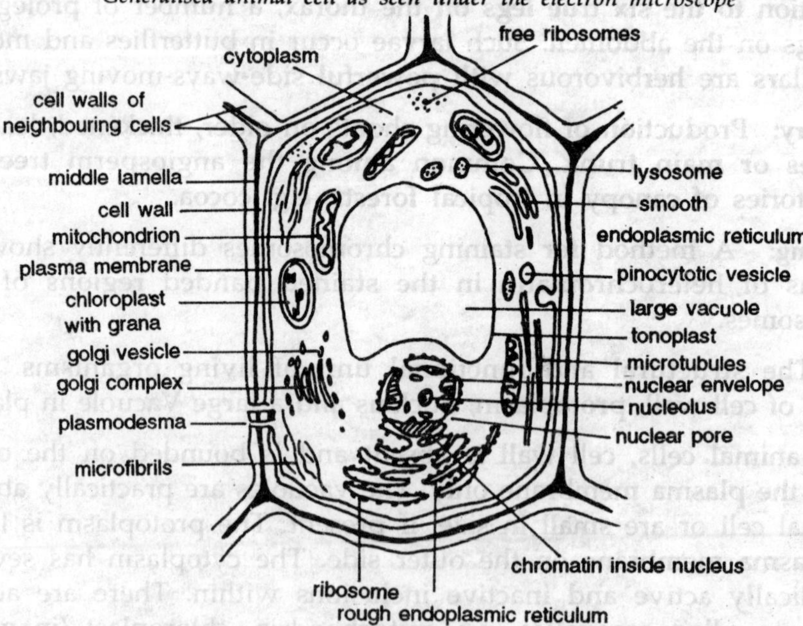

Generalised plant cell as seen under the electron microscope

Cell body: (Perikaryon). The mass of cytoplasm which contained nucleus and Nissl granules, from which the branches of a nerve cell arise.

Cell Cycle: Cycle of changes in the life span of a cell from the mitosis that forms the cell to the mitosis that divides it into daughter cells. The first phase after mitosis is 'G', when the cell has diploid quantity of DNA. Next phase of few hours is 'S' during which the quantity of DNA doubles up. It is followed by 'G when tetraploid DNA is present, Finally the mitotic phase 'M'.

Cell Division: Process by which single cell divides into two daughter cells.

Cell-mediated immunity: Specific immune response mediated by T-lymphocytes.

Cell membrane (Plasma membrane, plasmalemma): Extremely thin lipoprotein membrane Which covers all cells: It is a three-layered structure. It provides a selective barrier to the molecules and ions, Membranes of endoplasmic reticulum, mitochondria, etc.: have similar structure.

Cell Plate: Plate of materials appearing across the equatorial plane of the spindle during telophase in plant cells. Develops in middle lamella.

Cell sap: A watery solution of inorganic and organic substances present in the vacuole of a plant cell.

Cell theory: (i) All organisms are composed of cells and cell products, i.e., cell is the fundamental unit of structure of all organisms, (ii) All metabolic reactions in organisms take place in cells i.e., cell is the fundamental unit of function of all organisms, (iii) All cells arise from the division of preexisting cells, (iv) All cells contain 'hereditary substance which passes to the next generation by cell division.

Cell Wall: Outer Wall of plant cells. Comparatively rigid, giving mechanical support to plant tissue. In living cells traversed by extremely fine cytoplasmic threads the plasmodesmata which form delicate protoplasmic connections between adjacent cells. The walls of newly formed cells are at first very thin but as the cells assume their permanent character the walls thicken. Each cell has a primary wall consisting of cellulose, hemicelluloses and pectic substances. When cell has reached full size, it may remain with primary wall only, e.g. some forms of parenchyma. In other cells a secondary wall is laid down inside primary wall. It has a deposit of lignin.

Cellulase: An enzyme capable of digesting cellulose into simpler sugars. Associated with saprophytes and parasites, and particularly important when produced by the microflora in the herbivore's large intestine.

Cellulose: Fundamental constituent of cell wall, especially of higher plants, many algae, some fungi.

Cement: Bone-like substance which makes a thin covering to root of vertebrate tooth (i.e, below gum level) and in sonic mammals, ungulates) covers part of enamel of crown.

Cenozoic: The present era which extends from 65 million years ago. Age of mammals.

Central dogma: Doctrine that in all organisms, the genetic information flows from DNA to RNA and then to proteins.

Centre: A cluster of nerve cells that are concerned with a common function in the nervous system. Centres in the brain regulate functions as breathing, thirst, hunger, pain, pleasure, etc.

Centriole: Present in all organisms which have flagellated cells at some stage of their life. There are two small bodies composed of nine fused sets of microtubules, known as centrosome. Present near nuclear membrane of animal cell. Form astral rays and spindle fibres during cell division.

Centrolecithal: Pertains to eggs with yolk in the centre, as in arthropods.

Centromere: Region of chromosome for spindle attachment, also the site of attachment of sister chromatids.

Centrum: Massive part of each vertebra, lying ventral to spinal cord. Replaces embryonic notochord. Each centrum is firmly but flexibly attached to adjacent centra by collagen fibres.

Cephalochordata: A subphylum of chordates. Includes marine lancelets. These are small, fishlike chordates with notochord extending throughout body length even in the adult stage.

Cephalopoda: A class of phylum Mollusca, includes squids, cuttlefish, Octopus (devil fish) and Nautilus. These are marine softbodied animals with the head bearing a pair of large eyes. Giant squid, Architeuthis, is the largest invertebrate.

Cephalothorax: Fused head and thorax found in arachnida and many crustaceans.

Cercaria: Larval forms of flukes produced asexually by redia larvae while these are parasitic in snails. Ceicariae are infective to a new host to which they gain access either in blood (sheep live-fluke) or by penetrating skin. New host may be the definitive one in which the cercariae become sexually mature flukes or in some species it may be second intermediate host where they live for some time before being swallowed by definitive host and becoming sexually mature.

Cerci: A pair of filamentous appendages at the posterior tip of the abdomen on many insects and other arthropods.

Cereal: Fruit formed by plants of family Graminae. Used in food. For example, wheat, maize, rice.

Cerebellum: As subdivision of brain which lies behind the forebrain and above the brainstem. It controls the muscle movements for smoothly integrated body movements.

Cerebral: Pertaining to the brain as a whole or the anterior dorsal (cerebral) hemisphere.

Cerebral ganglion: Simple brain of many invertebrates.

Cerebral hemispheres: A pair of structures in vertebrate forebrain that contain the centres concerned with major senses, voluntary muscle activities, and higher brain function, such as language and memory. In mammals these are the largest part of brain and control the voluntary activities.

Cerebrospinal fluid: Fluid which fills the cavity inside vertebrate brain and spinal cord, and the piaarachnoid space outside thew., the inside and outside communicating by holes in the roof of the hind brain.

Cerebrum: The dorsal anterior part of the vertebrate forebrain consisting of two "hemispherical" masses.

Cerumen: The waxy secretion that collects in the external ear. Secreted by the ceruminous glands (modified sebaceous glands) to prevent the entry of dust particles, thus preventing damage of the tympanum.

Cervical: Pertaining to neck region. Cervical vertebrae are the vertebrae of neck.

Cervid: Member of family Cervidae (ruminants). For example, Dear, Moose,etc. These have antlers.

Cervix: Narrow posterior part of the uterus of mammals, which leads into the vagina.

Cestoda: **(Tapeworms)** A class of phylum Platyhelminthes including mostly endoparasitic forms with no digestive tract but with hooks and suckers for attachment and other parasitic adaptations and a complex life-cycle, e.g. *Taenia.*

Cetacea: An order of mammals, including whales, porpoises and dolphins. Totally aquatic with aquatic adaptations: no hair, blubber present, traces of pelvic girdle but no hind limbs while fore limbs modified as flippers. Blue whale (Balaenoptera) is the largest living animal.

Chaeta: A slender chitinous structure secreted by cells of the integument in many invertebrates like annelids for gripping. Also called seta or bristle.

Chaetognatha: Arrow-worms. A group of small marine animals sometimes included under the Phylum Annelida, e.g. *Sagitta.*

Chaetopoda: A group of annelids including bristle worm (Oligochaeta) and earthworm (Polychaeta). Well developed coelom.

Chalaza: (Bot) Basal region of Angiosperm ovule where stalk is attached. (Zool.) Of a bird's egg. Twisted strand of fibrous albumen. Two are attached to the vitelline membrane, at opposite poles of the yolk, lying in the long axis of the egg. They stabilize the position of the yolk and early embryo in the albumen.

Chalone: Substance secreted by a tissue that inhibits mitosis in that tissue.

Chamaephyte: A plant that perennates with its buds at or just above the soil surface.

Character displacement: Evolutionary divergence of two species that reduces the amount of niche overlap between them.

Chasmocleistogamic: Plants that produce chasmogamous (open) and cleistogamous (closed) flowers.

Chasmogamic: A flower that opens before pollination.

Chasmophyte: A plant which grows on rocks rooted in debris in the crevices.

Chela: Grasping appendage found in lobsters, crabs and other arthropods. Also called pincers.

Chelicerae: First pair of pincerlike appendages in spiders and other arachnid arthropods. These are associated with mouth and are formed for chewing or grasping.

Chelonia: Turtles and tortoises. Order of reptiles with body enclosed in plates of bone, usually covered with epidermal horny plates. Carapace is the upper while plastron is the lower part of their shell. Teeth absent.

Chemoautotrophic (Chemo-synthetic): Bacteria which obtain energy by oxidation of simple organic molecules; e.g.: the oxidation of hydrogen sulphide to sulphur by *Thiobacillus*.

Chemoreceptors: A sensory end-organ (afferent nerve endings or cells associated with them) which is capable of reacting to a chemical stimulus.

Chemotaxis: (Chemotactic movement) Movement in response to chemical concentration gradient. The spermatozoids of primitive plants are often positively chemotactic, swimming towards the female organs in response to a chemical secreted by the latter. For example, the archegonium (female organ) of the mosses secretes sucrose.

Chemotherapy: Treatment of disease by chemical substances which are toxic to the causal organisms.

Chemotrophic: Organisms obtaining energy by chemical reactions independent of light. These may be based on inorganic substances or organic substances from the environment.

Chemotropism: (Bot) Movement in which stimulus is a gradient of chemical concentration, e.g. downward growth of pollen tubes into stigma due to presence of sugars. (Zool.) Sometimes used synonymously with chemotaxis

Chiasma: Attachment of two nonsister chromatids in a bivalent in diplotene of prophase I of meiosis. Each chiasma results in the exchange of genetic material between the nonsister chromatids, i.e. in corssing-over.

Chilopoda: Group of arthropods under Myriapoda, including centipedes. Have a single pair of jointed appendages per segment, a pair of antenna and the first pair of appendages modified to form poison claws for catching the prey.

Chimera: Combination of tissues of different genetic constitution in same part of the plant.

Chiroptera: An order of mammals, highly specialised for flight. Includes bats.

Chitin: A horny protective substance forming the chief part of the cuticle of arthropods, an amino polysaccharide or a polymer of acetylglucosamine. It is insoluble in water, alcohol, ether and other solvents and is resistant to acids and alkalis.

Chlamydospore: A thick-walled asexual fungal spore which survives adverse conditions. It is intercalary and non-deciduous, being made up of one or more cells.

Chloranthy: An abnormal flower in which all the parts are developed as leafy structures.

Chlorenchyma: Thin-walled parenchymatous cells containing chloroplasts.

Chlorobacteriaceae: A: family of sulphur bacteria having a pigment like chlorophyll. They photosynthesize in presence of hydrogen sulphide.

Chlorocruorin: Respiratory pigment in blood of certain Polychaeta. A protein containing iron, Closely related to haemoglobin.

Chlorophyll: Green pigment found in all algae and higher plants (with exception of a few saprophytes and parasites). Located in chloroplasts except in blue-green algae (Cyanophyta) were it is borne on numerous photosynthetic membranes scattered in cytoplasm at cell periphery: raps sunlight during photosynthesis and converts it into chemical energy. It is of two types Chlorophyll a and Chlorophyll b.

Chlorophyta: Green algae. Possess chlorophylls a and b, alpha and beta carotene, xanthophylls, and store reserve food as starch, possess cellulose cell wall. Largest group of algae. Primitive form microscopic, unicellular, motile by flagella, or nonmotile; occurring singly or grouped into colonies. Higher forms multicellular with filamentous or flattened thallus.

Chloroplast: Chlorophyll containing plastids found in plants; largest cell organelle. Algal chloroplasts are variously shaped. Chloroplasts of their groups of plants are biconvex 5 to 10 mm by 2 to 4 mm in size. Double membraned. Internally there is an extensive double membraned lamellar system. In lower forms the pigments are evenly distributed over the entire surface of the thylakoids. In higher forms the thylakoids are closely packed as piles of coins in structures called grana. Usually 40 to 60 grana occur in a chloroplast. The grana contain granal thylakoids. Grana embedded in a matrix called stroma. 'Light' phase of photoynthesis occurs in the thylakoids whereas 'dark' phase occurs in the stroma.

Chlorosis: Yellowing of leaves due to deficiency of chlorophyll.

Choanae (Internal nares): Of vertebrates. Internal openings of nasal cavity into mouth.

Choanocytes (Collar cells): Flagellated collar cells in sponges. Maintain the flow of food-bearing water.

Cholesterol: A sterol (fat derivative) found in animals. Precursor of steroid hormones and bile acids. It occurs in bile, blood corpuscles, cell membranes, blood plasma, and egg yolk. It can accumulate in the gall bladder as gall stones, and is thought to be a contributory cause of coronary thrombosis at it is deposited in the walls of arteries causing them to narrow.

Choline: A constituent of B-complex group of vitamins. Its lack produces fatty liver in experimental animals. A constituent of acetylcholine.

Cholinergic: A nerve fibre secreting acetylcholine at its and when nerve impulses arrive there. In vertebrates, motor fibres to striped muscles, parasympathetic fibres connecting C.N.S. to sympathetic ganglia are all cholinergic.

Cholinestearase: An enzyme which inactivates acetylcholine.

Chondrichthyes: Cartilage fishes. True bone absent. Notochord present, throughout life. Placoid scales present. Air bladder absent. Examples Sharks and ray fishes.

Chondrocranium: The first part of the skull that form in vertebrate embryos. It consists of cartilaginous structures. It becomes ossified in adults, although it remains cartilaginous in Chondrichthyes.

Chondrocyte: A cartilage cell. It secretes the matrix of cartilage.

Chorda-mesoderm: Mesoderm and notochord of vertebrate embryo. In early stages of development, these tissues often form, a continuous mass or layer of similar cells, and they are closely related in physiology of development.

Chordata: Most highly evolved phylum of animals with notochord, gill slits and dorsal tubular single nerve cord, at least during development. Includes Protochordata (or Acrania) and vertebrates (Craniata).

Protochordata includes Hemichordata (Balanoglossus), Urochordata or Tunicata (Herdmania) and Cephalochordata (Amphioxus i.e. Branchiostoma).

Chorioallantois: A membrane formed by the fusion of the inner wall of the chorion and the outer wall of the allantois (chick).

Chorion: Embryonic membrane of amniote vertebrates, consisting of outer ectodermal epithelium with layer of mesoderm beneath. In mammals, the superficial layer enclosing all the embryonic structures. Its outer epithelium is trophoblast; forms placenta. (2) Noncellular membrane secreted around ovum by cells of ovary, e.g. the superficial envelope or 'shell' of an insect egg.

Chorionic gonadotrophin: The hormone secreted by the placenta which stimulates corpus luteum to release estrogen and progesterone.

Choroid: Middle layer of the vertebrate eve, between the sclerotic and the retina. Rich in blood vessels and contains pigment that absorbs light and stops it going through the eye past the retina.

Choroid Plexus: A highly vascular epithelial structure fining portions of the cerebral ventricles. Responsible for the formation of much of the cerebrospinal fluid.

Chresard: Total amount of water in the soil which is available to plants.

Chromaffin tissue: Tissue which produces adrenalin or noradrenalin. Occurs in adrenal medulla and is widely distributed in vertebrate body.

Chromatid: Each longitudinal half of chromosome. The two chromatids are visible during prophase and metaphase and separate during anaphase and are then termed chromosomes.

Chromatin: A nucleoprotein, staining with basic dyes, forming part of chromosomes.

Chromatophase (Bot.) Chromoplast: In procaryotes vesicles bearing photosynthetic pigments, (Zool.) Cells with pigment as granules in the cytoplasm, as in skin.

Chromocentre: Granules of condensed chromatin (heterochromatin) which are found scattered in interphase nuclei.

Chromomeres: Deeply staining granules found in prophase chromosomes during meiosis. Chromomeres in corresponding positions on homologous chromosomes pair during meiosis in many organisms. Probably mark positions where chromosome thread is more tightly coiled than in other regions.

Chromonema: An optically single thread forming an axial structure within each chromosome: The number of chromonemata (pl.) in a chromosome is found to be normally two.

Chromophore: Retinal component of a photopigment molecule.

Chromoplast: A plastid containing one or more coloured pigments, like red, yellow, orange, etc. For example, tomato fruits and carrot roots.

Chromosome: The physical basis of heredity. Deeply staining rodlike structures present in the nuclei of eukaryotes. Chemically nucleoproteins since they contain DNA and histories arranged in compact manner. Number is constant in cells of a particular species. They are replicated into identical chromatids during mitosis and meiosis.

Chromosome complement: The set of chromosomes characteristic of any cell of a species.

Chromosome map: Plan showing positions of the genes on a chromosome. For eukaryotes, it can be constructed from data of crossing-over, the frequency of crossing-over between the genes taken in pairs indicating their linear order, i.e. relative positions; the distances on such a map are given in units of crossover value. This provides a linkage map. Or it can be constructed by observations of localized aberrations of a chromosome (especially salivary) which produce phenotypes known to be associated with presence or absence of particular genes. The results of the two methods agree. Various methods of chromosome mapping are used: for prokaryotes; for bacteria, especially by the interruption at various times of the linear transfer of a chromosome between two cells during mating.

Chrysalis: The third stage in the development of a butterfly or moth, a pupa.

Chrysophyta: Golden-brown algae. Golden brown in colour due to abundance of carotenoid pigments, including a b carotene, fucoxanthin and other xanthophylls present with chlorophylla Many lack cell wall.

Chyle: The contents of the intestinal lymph vessels after a fatty meal.

Chyme: The partially digested food leaving the stomach.

Chymotrypsin: Proteolytic enzyme produced by pancreas which hydrolyzes peptide bonds in alkaline medium.

Cicada: Homopterous plant-bugs of warm regions having four membranous wings and long piercing mouth parts. The males, which live, only for a few days, can make a very loud rattling noise with drum-like membranes on the sides of the body., the females are silent.

Cicatryx: A thin, nonvascular side of the ovarian follicle where rupture occurs to allow the extrusion of the egg into body cavity in the vertebrates.

Cilliary body: Thickened circular rim of choroid of vertebrate eye, at border of cornea, containing ciliary muscles which produce accommodation. It secretes aqueous humour. To it the lens is attached and from it iris springs.

Cilliary feeding: Feeding in some invertebrates in which cilia, after creating a current of water towards the body, filter out food particles and transport them into the mouth.

Ciliata: A class of protozoans having cilia for locomotion; e.g. paramecium (slipper animalcule) and *Balantidium* (human endoparasite).

Ciliated epithelium: Sheet of cells, each cell with several cilia on exposed surface The cilia beat in coordinated rhythm. A common method of moving fluids in animal body, e.g. in respirator passages of land vertebrates,

Cilium: Fine cytoplasmic thread projecting from the surface of a cell. Cilia lash with orderly beat in a constant direction. Each cilium moves the fluid surrounding it by lashing through it like an oar and then bending on its recovery stroke so as to offer less resistance.

Circadian rhythm (diurinal rhythm): Endogenous rhythmic changes that occur in an organism with a periodicity of about twenty-four hours when the organism is isolated from daily rhythmical changes in its environment. Examples are leaf movements and growth in plants: sleep rhythms and running activity in animals. These are considered to be a fundamental expression of the ability of plants and animals to measure time although the underlying biophysical processes are largely unknown. Circadian rhythms are considered to be a basis of photoperiodism the two component phases of the circadian rhythm are postulated as scotophile (dark-loving) and photophile (light loving).

Circulatory System: System of passages and chambers through which materials are distributed in the body in blood or haemolymph. Small-sized simple animals secure distribution by diffusion of materials from cell to cell and thus have no circulatory system. In complex organisms, respiratory and excretory organs are far removed and hence rapid transportation is necessary. Round worms and onwards have a circulatory system which is a closed tubular circulatory system evolved first time in Annelida.

Circumnutation: Automatic growth movement in the growing tip of a plant showing curvature in a helical manner around the whole circumference.

Circumoral ring: Surrounding or around the mouth.

Cirrus: A small, slender and usually movable structure or appendage from a cell or body surface.

Cisternae: Flattened saclike vesicles of endoplasmic reticulum and Golgi apparatus.

Cis-trans effect: Two mutation at noncorresponding places somewhere in a given pair of homologous chromosomes (within the cells of a diploid organism) can be in either of two arrangements. (1) Both mutations may be on one chromosome of the pair, the homologous chromosome being normal (cis arrangement); or (2) each member of the pair carries one of the mutations, (trans arrangement). Similarly, two types of heterokaryons are possible, the cis heterokaryon where both mutations are carried in the same nucleus and the trans heterokaryon where the two mutations are in different nuclei.

Cistron: Gene defined functionally, i.e. the length of DNA producing RNA which in turn produces a specific Polypeptide and the trans: that functions in the economy of the cell.

Citric acid cycle (Krebs cycle): Also known as tricarboxylic acid cycle. Occurs during aerobic respiration within the matrix of mitochondria. The end-product of glycolysis-pyruvic acid is oxidatively de-carboxylated into acetyl coenzyme A. The latter diffuses into the mitochondria where it reacts with oxaloacetic acid in the presence of the enzyme citrate synthetase and a molecule of water to give rise to the six-carbon compound citric acid. The citric acid loses and then gains a molecule of water in the presence of the enzyme aconitase to produce its isomer isocitric acid. The latter is oxidized by isocitric acid dehydrogenase to produce oxalosuccinic acid, which loses a molecule of CO_2 to produce a five-carbon compound alpha ketoglutaric acid. It is oxidatively decarboxylated into a four-carbon compound, succinyl CoA which reacts with a molecule of water to produce succinic acid. A molecule of GTP is synthesized during the process. It is called substrate phosphorylation. Succinic acid is oxidized to fumaric acid in the presence of FAD containing enzyme. A molecule of water converts fumaric acid into malic acid which is oxidized to oxaloacetic acid in the presence of a dehydrogenase. The end-products of citric acid cycle are 3 molecules: of CO_2, 5 molecules of hydrogen and a molecule of ATP. The hydrogen molecules are subsequently transferred to oxygen in the process of oxidative phosphorylation to gate 3 ATPs per hydrogen.

Cladistic: In plants, reflecting recent origin from common ancestor.

Cladode: A modified internode of the stem, that functions as a leaf, being flattened and highly photosynthetic. It is a xerophytic adaptation and is seen in Asparagus. It is a phylloclade of single internode. Cladophyll. A flattened stem which functions as a leaf cladode.

Claspers: Any pair of processes on hind-end of the abdomen of a male insect, serving to grasp other structures during mating.

Class: A category in the taxonomic hierarchy, including one or more alike orders and itself is included along with its allies under Phylum. It is placed in between phylum and order; e.g. man belongs to the Class Mammalia which falls under the Phylum Chordata and in turn includes the order Primates for man.

Classification: 1. The grouping and arrangement of organisms into a hierarchal order, 2. The arrangement of organisms resulting from classification procedures. An important aspect of classification is their predictive value. For example if a characteristic is found in one member of a group of plants, then it Is also likely to be found in the other members of that group even though the characteristic in question was not used in the initial construction of the classification.

Clavicle (Collar bone): The bone which runs from the upper end of the breastbone towards the top of the shoulder a cross the root of neck. Supports the upper limb. Most frequently fractured bone in the body.

Claws: Most insects have two horny claws derived from the last segment of the tarsus. Generally there is a median pad or arolium between them.

Cleavage (segmentation): Repeated mitotic divisions in succession of the zygote cytoplasm, accompanying corresponding nuclear mitosis, which follows fertilization. In animals often produces a mass of small cells, the blastula. Usually called segmentation in plants. See holoblastic, meroblastic, bilateral cleavage, spiral cleavage.

Cleidoic egg: Egg of terrestrial animal enclosed within protective shell which largely isolates it from its surroundings, permitting only respiration and some loss (occasionally gain) of water(e.g. egg of bird or insect). Contrasted with most marine eggs, which exchange water and salts fairly freely with their surroundings.

Cleistogamy: Fertilization within an unopened flower. Such flowers never open.

Climatope: The climatic factors of a place constitute the climatope.

Climax: The final or stable community in a succession of natural plant communities in one area under a particular set of conditions. A climax is self-perpetuating and is in equilibrium with the physical and biotic environment.

Cline: A graded series of characters exhibited by a species or other related group of organisms, usually along a line of environmental or geographical transition. The populations at each end of cline may be substantially different from each other.

Clisere: Succession of climax communities in an area, each giving way to the next as a result of climatic changes.

Clitellum: Ring-like region of some Annelida (earthworms, leeches). Prominent in sexually mature animals, It contains glands which secrete a mucus sheath around copulating animals, thus binding them temporarily together, and a cocoon in which fertilization and development of eggs occurs.

Clitoris: The organ in the female mammal corresponding to the penis of the male. It is a small cylindrical organ situated in the angle of the vulva. An inch long, it is capable of erection during copulation.

Cloaca: Posterior part of the alimentary canal into which the urinary and reproductive ducts open, in birds, reptiles, amphibians and many fishes.

Clone: Descendants produced vegetatively or by apomixis from a single plant, asexually or by parthenogenesis from a single animal, by division from a single cell The members of a clone are of same genetic constitution.

Closed circulatory system: Circulatory system of higher organisms in which blood flows in a system of closed vessels (arteries, veins, capillaries) in the body.

Cnidaria: A subphylum of radially symmetrical diploblastic animal with tissue-grade of organization and stinging-cells or thread-cell present for offence, defence and feeding. Mouth is present, polymorphism (medusoid and polypoid zooids) seen in many examples also called Coelenterata, includes a free-swimming planula larva life-cycle.

Cnidocil: Trigger or a sensory conical projection on the free surface of the cnidoblast to receive chemical as well as physical stimuli from the surroundings.

CoA (Coenzyme A): Derivative of pantothenic acid (group B vitamin). It is, the coenzyme that transfers acetyl group from one reaction to another.

Coacervates: Fluid sediment, rich in colloidal substances having small globules from which early cell is supposed to have evolved.

Cobalamine (Vitamin B_{12}): Cobalt containing vitamin required by many organisms. Lack of it upsets cell division: In man, the haemopoietic principle necessary for red blood cell formation, is: produced from this vitamin by a substance secreted in the stomach.

Cocaine: Alkaloid which occurs in the cocoa plant. Used as a local anaesthetic.

Coccus: A spherical-shaped bacterium. Cocci may be found singly, in pairs (*Diplococcus*), or chains *(Streptococcus)*, or in regularly or irregularly packed clusters.

Coccyx: Lower end of spinal or vertebral column in man. Composed of 4 rudimentary small fused caudal vertebrae

Cochineal: A red dye extracted from the bodies of minute scale-insect Dactylopius coccus, which feeds on cactus plant.

Cochlea: Part of membranous labyrinth (inner ear) concerned in the' reception of sound with analysis of its pitch. A projection of the sacculus. In mammals coiled in a spiral.

Cocoon: A protective covering for the egg or more usually for the pupa of insects.

Cocaeine: Alkaloid which occurs in opium. Used in medicine as an analgesic, hypnotic, and in the treatment of coughs.

Co-dominance: The situation in which two different alleles are equally dominant. If they occur together the resulting phenotype is intermediate between the two respective homozygotes.: For example, if white antirrhinums (AA) are crossed with red antirrhinums (A'A'), the progeny (AA') will be pink. Sometimes one allele may be slightly more dominant than the other (partially or incomplete dominance) in which case the offspring, though still intermediate, will resemble one parent more than the other.

Codon: A sequence of three nucleotide bases (i.e. a nucleotide triplet) that codes for a specific amino acid. Since four different bases are found in nucleic acids, there are 64 (4 ´ 4 ´ 4) possible triplet combinations. The arrangement of bases along the DNA molecule synthesizes the genetic code. An average 150 triplets code for one polypeptide and when synthesis of a given protein is necessary the segment of DNA with the appropriate bases is copied in a molecule of messenger RNA. When the RNA migrates to the ribosomes, its string of codons is paired with the anticodons of transfer RNA molecules, each of which is carrying one of the amino acids necessary to make up the protein.

Coelenterata: Phylum of invertebrates, diploblastic and radially symmetrical. Single opening surrounded by tentacles serves both as mouth as well as anus. Coelom absent, single cavity coelenteron serves both as digestive tract and body cavity. Nematocysts present. Includes hydra and jelly fish.

Coelom: Body cavity of triploblastic animals in which gut is suspended, lined entirely by mesoderm, especially peritoneum.

Coenobium: Colony of algal cells, constant in number and arranged in a specific manner. It coordinates and behaves as a unit. For example Volvox.

Coelomoduct: Cilliated duct connecting the coelom with external environment. Coelomoducts provide a means of exit for gametes and waste products.

Coenospecies: A group of species related by the possibility of producing fertile hybrids with one another.

Coenocytic: Multinucleate condition in many plants where there are many nuclei in a continuous cytoplasm.

Coenzyme: A nonprotein organic group without which certain enzymes are not active. The protein part of an enzyme is known as the apoenzyme and when united with a coenzyme, either permanently or temporarily, the two form an active enzyme known as a holoenzyme.

Cofactor A nonprotein substance that helps an enzyme to carry put its activity. Cofactors may be cations or organic molecules, known as coenzymes. Unlike enzymes, they are, in general, stable to heat, when a catalytically active enzyme forms a complex with a cofactor, a holoenzyme is produced. An enzyme without its cofactor is termed an apoenzyme.

Cohesion: Force of attraction between like molecules.

Coitus: Copulation of male and female, a term generally used in connection with mammals.

Colchicine: Drug (an alkaloid) which prevents mitosis proceeding beyond metaphase (by inhibiting spindle through disruption of microtubules). In actively dividing tissue of an organism treated with colchicine there is an accumulation of cells arrested in metaphase making easy the detection of proliferating regions, Reversion of colchicine-inhibited metaphase nuclei to the resting state results in polyploidy since the chromosomes have duplicated without the chromatids separating into daughter nuclei. This is a method of inducing polyploidy artificially, important in obtaining new agricultural and horticultural varieties.

Coleoptile: Protective sheath surrounding plumule in monocot seedling.

Coleorhiza: Protective sheath surrounding radical in monocot seedling.

Collagen: Fibrous protein which on boiling yields gelatin, Forms intercellular fibres ('white fibres'of connective tissue). Fibres made of collagen have a high tensile strength (tendon) but unlike elastin fibres have little reversible extensibility.

Collar cell: Flagellated or choanocyte cell bearing a flagellum and peculiar of sponges bounding some of the internal cavities called radial canals. These produce and maintain water current in the canal system.

Collateral vascular bundle: Vascular bundle in which the phloem is external (on one side) to the xylem and on the same radius.

Collecting duct: Portion of the tubular component of a kidney nephron. Ducts from several nephrons combine and empty into renal pelvis.

Collenchyma: Supporting tissue composed of living cells with the primary walls thickened at the corners of the cells.

Colloidal solution: Solution in which the solute is present in the colloidal state.

Colloidal State: System of particles in a dispersion medium. Differs from a true solution because of the larger size of the particles. As a result of the grouping of the molecules a solute in the colloidal state cannot pass through a semipermeable membrane.

Colon: Large intestine of vertebrates, excluding the narrower terminal rectum. In amniotes and some Amphibia, but not in fish, clearly marked off from small intestine by valve. Largely concerned with absorption of water from faeces.

Colony: A group of organisms of the same species living together; colonial. Opposite of solitary.

Colostrum: Maternal milk of mammal formed during the first few days after the birth. Particularly rich in proteins, including antibodies in some animals, especially ungulates.

Colour blindness: (Red-green). Genetically controlled defect in which one cannot distinguish between red and green colour.

Columella: (1) Dome-shaped structure present in sporangia of many phycomycete fungi of the order Mucorales produced by formation of convex septum cutting off sporangium from hypha bearing it. (2) Sterile central tissue of moss capsule.

Columella auris: Rod of bone or cartilage connecting eardrum to inner ear and transmitting sound in reptiles, birds and Anura.

Commensalism: A relationship between an organism and its host in which the host neither benefits nor suffers from the association.

Commissural fibres: Nerve fibres connecting one lateral half of the brain or spinal cord with the other.

Commissure: Transverse cord of nerve-fibres linking members of each pair of ganglia in double nerve-cord of Arthropoda and Annelida. Circum-oesophageal commissure, nerve cord connecting ganglia in head above oesophagus with those below it in e.g. Arthropoda and Annelida. Bundle of nerve-fibres connecting right and left sides of brain or spinal cord in the vertebrates. Nerve cords uniting ganglia of Molluscs.

Community: Any naturally occurring group of different organisms inhabiting a common environment, interacting with each other especially through food relationships, and relatively independent of other groups. Communities may be of varying sizes, and larger ones may contain smaller ones.

Companion cells: Small cells having dense cytoplasm and prominent nucleus lying side by side with sieve-tube cells in phloem of flowering plants.

Compensatory hypertrophy: Increase in size and functioning of residual part of a tissue or organ, some of which has been removed or put out of action, e.g. when one kidney is removed, other one enlarges.

Competition: Utilization of the same resources by one or more organisms of the same or of different species living together in a community, when the resources are not sufficient to fill the needs of all the organisms.

Complementary genes: Genes that can only be expressed in the presence of other genes. For example, if one gene controls, the formation of a pigment precursor and another gene controls the transformation of that precursor into the pigment, then both genes must be present for the colour to develop in the phenotype. Such interactions between genes lead to apparent deviations from the 9:13:1 dihybrid ratio in the F_2. For example, if two complementary genes control a, certain character, the dominant alleles of each of the two genes must be present for the character to appear, then a 9:7 ratio is seen.

Complementation: (1) Inter-cistron complementation. The effect of two homologous chromosomes, each mutant and defective but in different cistrons, which when together in a diploid organism in a heterokaryon, supply each other's deficiencies and give a Phenotype which is very similar to, or identical with, normal (wild type). (2) Intra-cistron (or interallele) complementation. Interalleles of a diploid organism or in a heterokaryon of two alleles of the same cistron give a phenotype closer to, but not identical with normal (wild type) than either could give alone.

Compound eyes: Eyes of insects and crustaceans consisting of numerous visual units or ommatidia. When crowded together the facets are hexagonal, but are circular when there are only a few. Each ommatidium consists of a cuticular lens beneath which is a crystalline cone and a retinula or group of light-sensitive cells. The whole eye is convex or hemispherical with the apices of the cones converging towards the optic nerve in the centre.

Complementary nucleotides Nucleotides which pair specifically with each other.

Concentric Vascular bundle: Vascular bundle with either phloem surrounding the xylem or xylem surrounding the phloem.

Conceptacle: Cavity containing sex organs, occuring in groups on terminal parts of branches of thallus in some brown algae, e.g. Fucus.

Concrescence: The coming together of previously separate parts (cells) of the embryo generally resulting in pilling up of parts.

Conditioned reflex: Induced reflex: reflex modified by experience, the original sensory component (i.e., the stimulus, sense organ and sensory nerve path involved in it) being replaced by a different sensory component, the motor component (i.e., the response remaining unchanged). Food placed in a dog's mouth evokes reflex flow of saliva. If the introduction of food is accompanied by ringing a bell, and this is repeated several times, the bell alone eventually becomes able to evoke the salivation, though the effect is temporary unless reinforced from time to time by administering both food and bell. The original salivation reflex, which is inborn, i.e. does not depend on the experience of the animal, is the unconditioned reflex; it is not lost as a result of the experiment. The induced response to the bell alone is a conditioned reflex.

Condyle: Knob-like part of bone which fits into a corresponding socket of another bone. The condyle and its socket form a joint allowing movement in one or two planes but no rotation; e.g. condyle at each side of lower jaw where it articulates with skull; occipital condyles on skull of tetrapods fitting into atlas vertebra.

Cone: (Bot) (Strobilus). Reproductive structure consisting of a number of sporophylls more or less compactly grouped on a central axis, e.g. cone of pine tree. (Zool). Kind of light-sensitive nerve cell present in retina of most vertebrates (though usually not in those which live in dim light) The cone shaped outer segment of the cell, which is the light sensitive part, consists largely of a stack of flat and parallel unit membranes at right angles to the cell length; it develops embryologically from a cillium and retains its characteristic 9 plus 2 pattern of fibrils. There are about 6 million cones in retina of a primate. Concerned in colour discrimination, and in the most acute discrimination of detail.

Confluence: Similar to conerescence except that this term refers specifically to the 'flow' of cells (areas) together, whether or not they are piled up.

Congenital: Deformity that exists at or before birth.

Conglobate gland: Phallic gland in male cockroach, large saclike opening near the male aperture and secreting the outer most lining of the spermatophore.

Conidium: **(Conidiospore)** An asexual spore of certain fungi; e.g. *Pythium* and *Albugo*. They are cut off externally in chains at the apex of a specialized hypha, the conidiophore.

Coniferales: Order of Gymnospermae including extinct forms and nearly all living gymnospermspine, spruce, cedar, yew, larch, etc., characteristic of temperate regions; majority tall, evergreen, forest trees. Usually monoecious with distinct male and female cones. Microsporophylls borne directly on cone axis female cones compound, ovules borne on ovuliferous scales in the axils of bracts arising from the cone axis.

Conjugated protein: A simple protein combined with some other compound, such as a carbohydrate (glycoprotein) or pigment (chromoprotein).

Conjugation: (1) Union of two paramecia (ciliate protozoan) for the exchange of their genetic materials. (2) Union of gametes, free-swimming or not, especially in isogamy. (3) Union between two cells in certain bacteria (*Escherichia* and related genera) allowing passage of genetic material from one, the donor cell, to the other, the recipient cell. One important difference from sexual reproduction in other organisms ventral side. Meets scapula at glenoid cavity. A cartilage-bone. Reduced to small process of scapula in mammals.

Coral: The hard deposits built up by minute colonial anthozoan coelenterates. Most corals are actinozoans except stinging coral (fire coral) or *Millepora* and *Stylaster*, etc. which are hydrozoans, while most anthozoans are corals except sea-anemones, etc. These occur chiefly in the warmer seas and they may be solitary or colonial, and many times make coral reefs of the fringing (shore), barrier and circular (atolls) types. Many other animals and plants help in this formation and they with others live in the reefs, e.g. red coral (*Corralium*).

Corium: The dermis portion of the skin beneath the epidermis.

Corm: A swollen, underground stem with buds in the axil of scale leaves. It has adventitious roots and is responsible for vegetative reproduction.

Cormophytes: Plants possessing stem, leaf and root i.e. spermatophyta.

Cornea: Transparent part of the sclerotic at front surface of eye of vertebrates, overlying iris and lens. Mainly responsible in land vertebrates for refraction which results in focussing of image on retina.

Cornification: Conversion of cell content into dead keratin. For example, in epidermis of skin.

Corolla: Coloured part of a flower, within the calyx; consists of a number of petals.

Coronary occlusion (coronary thromobosis): Obstruction of a coronary artery or one of its branches that supply blood to the bean muscles. It results in sudden death and it is due to arteriosclerosis and hypertension: Coronary vessels. Arteries and veins of vertebrates which supply blood to heart muscles.

Corpora allata: A pair of small ductless glands behind the brain in insects and their larvae. These secrete juvenile hormone (neotenin) which maintains the insect in larval state at each moult. When production of this hormone ceases, metamorphosis takes place.

Corpora quadrigemina: Four little rounded elevations on the dorsal surface of midbrain.

Corpus callosum: A broad tract of nerve fibres within the brain of marsupial and placental mammals, connecting the two cerebral hemispheres.

Corpus luteum: Temporary organ of internal secretion, of the hormone progesterone. Formed in mammals in interior of a ruptured Graafian follicle after ovulation, by ingrowth of follicle wall, which becomes secretory luteal tissue. If ovulation does not result in fertilization, then corpus luteum soon degenerates. If fertilization occurs, corpus luteum persists and continues secreting during part or all of pregnancy.

Corticotropin releasing factor: A chemical agent from the hypothalamus which stimulates the release of ACTH by the adeno-hypophysis.

Cortisone: One of the hormones produced by cortex of adrenal gland; predominantly a glucocorticoid: Has usual complex effects of cortical hormones, and notably produces a diminution of local inflammation and healing response.

Corymb: A raceme inflorescence with flowers borne at the same level due to elongation of pedicels of lower flowers. For example, candytuft.

Cosmoid scale: Scale typical of primitive Choanichthyes.

Costal: Pertaining to the ribs. Intercostal refers to the space between the ribs.

Cotyledon: Leaf forming part of embryo of seeds; much simpler in structure than later formed leaves and usually lacking chlorophyll. Monocotyledons have one, dicotyledons have two cotyledons in each seed. Play an important part in the early stages of seedling development. In seeds like peas, beans, they are storage organs from which the seedling draws food. In seeds like grasses, food stored in the endosperm is absorbed by the cotyledons and passed on to the seedling.

Courtship: Special behaviour of animals in seeking mates. Varies from complex behaviour of birds and mammals to the random association of the sexes in many simpler animals.

Cowper's gland: A gland of male reproductive ducts of mammals whose acidic secretion is a part of the seminal fluid. This secretion kills bacteria etc. in the urethra, and is neutralized in the vagina during copulation by the bertholin glands secretion (basic).

Coxa: The basal segment of an arthropod leg with which the leg is joined on the ventral surface of the body.

Coxal gland: Excretory glands opening on the fifth segment of the body in king crab, scorpions and spiders (Arachnida), and in some insects.

Cranial: Pertaining to the skull or brain, as a cranial nerve.

Cranial nerve: Peripheral nerves emerging from brain of vertebrates. Dorsal and ventral roots of several segments exist amongst these nerves, but (unlike those of spinal cord) remain separate. Each root is numbered and named as a separate nerve. There are ten pairs of cranial nerves on each side in living anamniotes, eleven or twelve in amniotes and fossil amphibians. Include nerves supplying eyes, ears, nose, jaw muscles, skin of face, etc., e.g., olfactory, optic, oculomotor, trochlear, trigeminal, abducens, facial, auditory, glossopharyngeal, vagus, accessory and hypoglossal nerves,

Cranium: Skull of vertebrates: Encloses brain.

Creatine phosphate: An energy rich compound especially prevalent In muscle tissue.

Crenation: The shriveling of red blood cells due to withdrawal of water.

Cretaceous: Geological period lasting approximately from 35 till 65 million years ago.

Cretinism: Abnormal condition resulting from the underactivity of thyroid in a child. It leads to mental retardation and underdeveloped body.

Crinoidea: A class of echinoderms, including sea-lillies with body attached by a stalk; mouth and anus on oral side; 5 long arms with pinnules; tube feet without suckers, no spines, pedicellaria and madreporite, e.g. *Antedon*. Larva is a fixed cystidean or pentacrinoid.

Crop: An enlarged and modified part of the oesophagus where food may be kept for a time before passing on to the gizzard or stomach as in insects, leeches and birds, etc.

Crop milk (Pigeon's milk): Secretion consisting of crop epithelium of mate and female pigeons on which nestlings are fed. Like mammalian milk, its production is influenced by hormone prolactin.

Cross: Act of cross-fertilization, or the organism resulting from cross-fertilization.

Cross-fertilization: Union of an egg cell from one individual with a sperm cell from another individual of the same species.

Crossing-over: The exchange of material between homologous chromatids by the formation of chiasmata during meiosis.

Cross-Over Value: (C.O.V.): Frequency of crossing over between two genes in different parts of the same chromosome, expressed as percentage of gametes in which one of the two genes has been exchanged for an allele from the homologous chromosome.

Cross-pollination: Transfer of pollen grains to the stigmas of the flowers on different plants of the same species: There is mixing of dissimilar characteristics.

Crustacea: A class of arthropoda with more than 4 pairs of legs; body divided into cephalothorax and abdomen; compound eyes, respiration by' gills; includes prawns lobsters, shrimps, crabs, crayfish, barnacles and hermit crabs, etc.

Cryophytes: Plants growing on ice and snow. Micro plants; largely consisting of algae and also some mosses, fungi and bacteria. Algal forms may be so abundant as to colour substratum; e.g. 'red snow' due to the presence of *Chlamydomonas.*

Cryptic coloration: Coloration of animals which conceals by resemblance to the surroundings.

Cryptogam: In early classifications, any plant that reproduces by spores or gametes rather than by seeds. They include the algae, fungi, bryophytes, and pteridophytes.

Cryptozoic: Animals inhabiting crevices, e.g. under stones, leaves.

Ctenoid scales: Hard scales of fishes having rough or comblike edge.

Ctenophora (Acnidaria): Includes comb-jellies or comb-girdles or sea walnuts. Similar to coelenterates in all respects except for biradial symmetry, 8 ciliated comb-plates externally present for locomotion, stinging cells absent and mesoglea more, thick. This minor phylum includes venus girdle (*Cestum, Velamen*), etc.

Cultivar: Any agricultural or horticultural 'variety'. The term is derived from the words 'cultivated variety'.

Culture medium: Mixture of nutrients, which may be used in liquid form or solidified with agar; used to cultivate microorganisms such as bacteria or fungi, of support tissue cultures.

Cumulus oophorus: The cell aggregation which immediately surrounds the mammalian egg within its Graafian follicle.

Cupule: 1. A cup-shaped structure in Which certain fruits (e.g. a corm and hazelnut) are borne. 2. A protective cup made up of six modified leaves surrounding the young gemma of *Lycopodium*. 3. The bright red tissue surrounding the ovule of Taxus, 4. The ovule-bearing structure terminating the pinna in the extinct Caytoniales.

Cutaneous respiration: Respiration from the surface of skin, as in frog,

Cuticle: A noncellular layer secreted by the epidermis of an animal.

Cutin: Waxy substance secreted by plant epidermis and forms cuticle.

Cyanophyta: (Blue-green algae, Myxophyta). The division comprising the prokaryotic algae. They are extremely simple plants with little internal differentiation and resemble photosynthetic bacteria in many features. There are no chloroplasts or mitochondria and the photosynthetic and respiratory mechanisms are thought to be located on the internal lamella system. Food is stored as glycogen and cyanophycin, and reproduction is asexual, The cyanophyta have little in common with other algae except for their photosynthetic pigments. Examples are *Nostoc* and *Oscillatoria,* which are nitrogen-fixing, while some others are partners in lichen associations.

Cygadales: Order of Gymnospermae comprising fossil (Mesozoic) and a few living representatives. Most primitive living seed plants. Indigenous to tropical and subtropical zones. Stem unbranched, tuberous or columnar, upto 20 metres in height, bearing a crown of fernlike leaves. Dioecious. Microsporophylls distinct from vegetative leaves, arranged in a compact cone. Megasporophylls, leaflike structures, loosely grouped, or highly modified, grouped in a compact cone. Male gametes motile by cilia.

Cycadorilicales (Pteridos-permae): Order of extinct, palaeozoic Gymnospermae that flourished particularly during the Carboniferous: of great phylogenetic interest: Reproducing by seeds but possessing fern like vascular system with development of secondary wood: Micro-and megasporophylls little different from ordinary vegetative fronds; not arranged in cones.

Cyclamates: The artificial sweetening agents which may be carcinogenic.

Cyclic AMP: Form of adenosine monophosphate produced by the enzyme adenyl cyclase at the cell surface from ATP, and diffusing thence into the interior of the cell where it activates various enzymes. The production of cyclic AMP is known to occur in a number of cells sensitive to particular hormones when the hormone impinges on the cell; and it is an essential link in the chain of reactions set going by the hormone' AMP seem also to be concerned in the release of the neutral transmitters, adrenaline and noradrenaline, when an impulse arrives at an adrenergic ending.

Cyclops: A genus of minute crustaceans found universally in fresh water and an important member of the fresh water plankton.

Cyclostomata: A class of most primitive living vertebrates of seas with fishlike body but without paired fins, vertebral column *indistinct;* suctorial mouth for feeding on fishes; skin without scales and mouth without true jaws.

Cymose inflorescence (Cyme, Definite inflorescence): An inflorescence in which apical growth is terminated by the formation of a flower at the apex. Subsequent growth is then from lateral buds below the apex, which themselves form flowers and more lateral shoots.

Cypsela: Fruit of compositae family similar to achene but is derived from an inferior ovary.

Cyst: A resistant protective covering formed around a protozoan or other small organism during unfavourable conditions or reproduction; a small sac or capsule.

Cysticerous (Bladder worm): Infective larval stage of tape worm in its primary host, consisting of a scolex tucked into a large bladder; occurs in intermediate host; grows into mature tapeworm, if swallowed by definitive host. Pork infested with bladder worms is called measly pork.

D

Damping off: Disease which attacks young seedling at ground level causing them to rot and fall. Usually caused by the fungus *Pythium.*

Dark reaction: Takes place during photosynthetic process. Carbon dioxide is fixed to form sugars by combination with hydrogen. Takes place in the stroma of the chloroplast.

Darwinism: Darwin suggested that in many cases the differences may be harmful to the organism, making it less equipped to fit its environment and reducing its ability to compete with members of the same species. On the other hand, an inheritable change may occur that gives an organism a better chance of survival. This individual will pass on its advantageous characteristics to its offspring which will stand a higher chance than normal of surviving, breeding, and producing further altered descendents. They eventually out number the unalteted organisms. The change gives an organism a better chance of survival depending largely on the natural environment. Darwin, therefore called this process natural selection.

Dauer modification: A lasting inheritable change, possibly cytoplasmic, which is produced by some treatment.

Day neutral plants: Plants in which flowering is not affected by length of photoperiod. For example, tomato.

Day vision: The vision in bright light, in contrast to night or twilight vision.

D.D.T: Dichlorodiphenyl trichloroethane; used as an insecticide.

Dead space: The amount of space in the respiratory tract in which none or very little respiration takes place (the nasal passage, trachea, bronchi and bronchioles).

Deaminase: Enzyme which catalyses the removal of an amino-group from a compound.

De-amination: Removal of amino (NH_2) group, especially from an amino acid.

Death: The termination of vital processes in the organism. Only organisms that reproduce by binary fission are immune from normal death.

Decapoda: 1 A large and important order of Crustacea (Arthropoda) including shrimps, prawns, lobsters, crayfishes and crabs. 2. An order of Cephalopoda (Mollusca), including cuttle fish and squids.

Decarboxylation: The removal of carbon dioxide from the carboxyl group of an organic acid.

Decidua: Mucous membrane (endometrium) lining uterus in the thickened and modified form it acquires during pregnancy in many mammals. Some or all of the decidua comes away with the placenta at birth.

Deciduous: To fall off or shed: some trees shed leaves during autumn.

Deciduous teeth (Milk teeth): First of the two sets of teeth that most mammals have; similar to the second (permanent) set which replaces it, except in having grinding teeth corresponding only to the premolars, not to the molars, of that set.

Decomposers: Organism that utilizes dead plant or animal material for food and releases the component elements to the environment, thus contributing to circulation of these elements in the ecosystem.

Decompound: When in a compound leaf, the leaflets are made up of distinct parts.

Decumbent: Lying flat or creeping on ground, but having the tip growing upwards.

Decussate: Said of leaves, when they are opposite, but each pair is at right angle to the ones above and below it.

Dedifferentiation: Reversion of a specialised cell to a more generalised, embryonic type. Occurs during initial phase of regeneration of salamandar limbs.

Defecate: To discharge faeces through anus.

Deficiency: Loss of a terminal acentric segment of a chromosome.

Deficiency disease: Disease caused by deficiency or absence of some essential food substances. e.g. vitamins, minerals or essential amino acids.

Definite: Always of the same number in a given species, Of a stem, ending in a flower, and so stopping growth. Of a stem, when the bud grows rapidly to its full length and then stops. Said of an inflorescence when all its branches end in flowers.

Deflocculation: Aggregation of clay particles: This forms a sticky mass, which is difficult to redisperse, thus destroying soil-structure.

Defoliant: Chemical which induces premature leaf-fall.

Deglutition: Swallowing, a complex reflex set off by stimulation of pharynx involving contraction of muscles of mouth and pharynx, closing of the back of the nose by soft palate, raising of larynx against epiglottis, inhibition of breathing, peristalsis in oesophagus,

Degraded ecosystem: (Sometimes called a Disclimax). An ecosystem of fluctuating communities resulting from recent major disturbance and containing few trophic levels. It has a less complex food web than a natural ecosystem.

Dehiscent: Opening spontaneously to release spores or seeds.

Dehydration: The excessive loss of water from the tissues of the body.

Dehydrogenase: Enzyme which catalyses oxidation reactions by removing hydrogen from the substrate. The hydrogen may combine with molecular oxygen or with another substance. Most oxidizing enzymes are dehydrogenases.

Delamination: Separation of cell layers by splitting, a process in mesoderm formation.

Deletion: Loss of an intercalary acentric segment of a chromosome.

Deme: Basic unit in terminology of categories in experimental taxonomy. Any group of individuals of a specified taxon of plants possessing clearly definable genetical, cytological or other characteristics, e.g., gamodeme, group of individuals which are capable of interbreeding.

Demerbal: Fish or other animals living near bottom of sea.

Demography: Numerical study of human populations.

Denaturation: Structural change produced in soluble (globular) proteins by mild heat or various chemicals rendering them less soluble than in their original ('native') condition, involving an unfolding of the peptide chains. Alteration of the physical properties and three-dimensional structure of a protein by agents too mild to break the peptide bonds. Of DNA, separation of the two strands, as by heating.

Dendrite: Branching, usually short, extension of a neuron which conducts impulse to the cell body.

Dendrograph: An instrument which is used to measure the periodic shrinking and swelling of tree trunks.

Dendrology: Branch of botany concerned with trees and shrubs.

Dengue: "Breakbone fever" or "dandy fever", An acute fever due to a filterable virus transmitted by female *Aedes aegypti* mosquito; characterized by sudden onset of an acute fever with chill, severe headache, marked pain in the muscles and joints and profuse sweating; no specific treatment.

Denitrification: The decomposition of nitrates in the soil by *Bacterium denitrificans,* resulting in the production of nitrogen.

Denitrifying bacteria: Soil bacteria which break down nitrites and nitrates anaerobically to produce free nitrogen.

Dental formula: Formula indicating, for a given species of mammal, the number of each kind of its teeth. The number in the upper jaw of one side is written above that in the lower jaw of one side; and the categories are given in the order incisors, canines, premolars, molars. The formula of the primitive mammal is i 3/3. c 1/1. pm 4/4. m 4/3. The human formula is i 2/2. c 1/1. pm 2/2. m 3/3.

Dentary: A dermal bone; one of several lower jaw bones of vertebrates but the only bone forming lower jaw in mammals.

Dentate: Having a toothed margin.

Dentine: Main constituent of teeth. Like bone in structure, but contains no cells, though cell processes penetrate it from cells in pulp cavity of tooth. Lies beneath enamel. Formed from mesoderm of neural crest of embryo. Ivory is dentine.

Dentition: Form and arrangement of teeth in the vertebrates. Teeth primitive in shark, with sharp flattened point and arranged in several rows, replaceably; in rest of the fishes and amphibians, numerous and sometimes vomerine also; in reptiles and mammals, a row of teeth along each jaw, entirely conical in reptiles but of various types in mammals, may be indefinitely renewable or limited to one or two sets; absent in birds.

Deoxyribonucleotides: Pro-ducts of hydrolysis of DNA.

Deoxyribo nucleic acid (DNA) The hereditary material of the cell. The molecule contains deoxyribose (a sugar), phosphoric acid, two pyrimidines (thymine and cytosine) and two purines (adenine and guanine). The pyrimidines and purines are bases. The molecule is a double helix of sugar-phosphate linkages (forming the sides of a twisted ladder) with the bases joined across to form the rungs. Adenine and thymine are always paired, as are guanine and cytosine.

Deoxyribose: The carbon sugar (pentose), having one oxygen atom less than the parent sugar ribose; component sugar of DNA.

Depolarize: To change the membrane potential towards zero so that the inside of the cell becomes less negative

Dermal: Pertaining to skin especially the inner connective tissue layer of vertebrate skin.

Dermal bone (Membrane bone): Bone arising near surface of embryo from concentration of mesenchyme cells, with no cartilage precursor. Occurs in skull and shoulder girdle of most vertebrates, e.g., roofing bones of skull; dentary clavicle.

Dermaptera: Earwigs, Hemimetabolcus insects in which the forewings are reduced to small, thickened elytra and the membranous, hindwings are roughly semicircular. 'There is a pair of large forceps at the hind end of the abdomen.

Dermatogen: The external layer of a stein or root apex one cell thick, and giving rise to the epidermis.

Dermatology: The study, diagnosis and treatment of the skin and its diseases.

Dermis (Corium): Innermost of the two layers of the skin of vertebrates, the outer being the epidermis. Much thicker than epidermis. Consists of connective tissue, with abundant collagen fibres mainly parallel to surface; scattered cells; blood and lymph vessels; sensory nerves. Sweat glands and hair-follicles project down from epidermis into dermis. Responsible for tensile strength of skin. May contain scales or bone. Scales dermal in origin in fishes.

Dermoptera: An order of mammals containing only the flying lemurs of the Malayan region; differ from mammals in dentition: like flying squirrels; have lateral folds of skin which support them in air.

Descending aestivation: Aestivation in which each segment overlaps the one anterior to it.

Desmolases: Enzymes breaking, or forming a C-C link without hydrolysis, e.g., carboxylase catalyses the change of pyruvic acid to acetaldehyde and carbon dioxide.

Dermosomes: Attachments between adjacent cells.

Determinant (Antigenic): The part of an antigen molecule that actually forms a specific bond with a corresponding antibody molecule. (Genetics) A factor that transmits inheritance, either a gene or a plasmagene.

Detritous food chain: The food chain in which microorganisms absorb and break down the energy rich compounds synthesized by the primary producers.

Detritus: Organic debris from decomposing plants and animals

Deuteromycetes: Fungi imperfecti. Fungi where sexual stages are not known.

Deutoplasm: Yolk, vitellin or secondary food substance of the egg; nonliving, resisting cleavage.

Devil rish: Octopus (Mollusca: Cephalopoda). Has 8 arms.

Devonian: The 'Age of Fishes', some 405-335 million years ago, between the Ordovician and the Carboniferous periods of number and variety of fishes, most of which have become extinct without leaving any modern relatives. During the late Devonian, primitive amphibians were evolving from crossoptery-gians (lobed finned fish). Vascular land plants appeared, such as the psilophytes and pteridophytes, while terres-trial fauna included insects and spiders. The period is named after rocks found in Devon.

Diabetes insipidus: A metabolic disorder marked by thirst and the passage of large amount of urine and without excess excretion of sugar. Due to defective ADH control.

Diabetes mellitus: A disease in which hormonal control of plasma glucose is defective because of deficiency of insulin. It is marked by glucose excretion in urine and high blood-glucose level.

Diadelphous: Stamens united by their filaments to form two bundles.

Diageotropism: Orientation of plant part by growth curvature in response to stimulus of gravity so that its axis is at rightangles, to direction of gravitational force, i.e. horizontal.

Dialysis: A technique for separating compounds with small molecules from compounds with large mole-cules by selective diffusion through a semipermeable membrane. For example, a mixed solution of starch (large molecules) and glucose (small molecules) is placed in a bag or piece of tubing made of thin cellophane or other suitable material. If the container is put in water, the sugar molecules diffuse out leaving the starch behind. Dialysis is performed naturally by the kidneys to extract wastes from the blood, or in the case of lost or damaged kidneys artificially by machine.

Diandrous: Having two antheridia, or two stamens.

Diapause: A slowing down of metabolism and delay of development of insects etc. Orten connected with seaso-nal changes of atmospheric conditions.

Diapedesis: Passage of a blood corpuscle through vessel walls without rupture. Squeezing of the leucocytes out of the capillary walls.

Diaphragm: A sheet of tissue, part muscle, part tendon, covered by serous membrane, separating cavities of thorax from cavity of abdomen. Present only in mammals. Helps in breathing: Diaphragm is arched up into thorax at rest and its flattening is a very important part of mechanism of inspiration or breath in most mammals.

Diaphysis: Shaft of a long limb-bone, or central portion of a vertebra, in mammals.

Diarrhea: Increased frequency of bowl movements, the stool tending to he liquid.

Diastase: An enzyme complex which breaks down starch to glucose.

Diastema: Space created on the jaw by the absence of canines as in the orders Lagomorpha (rabbit and hare), Rodentia (rat and squirrel) and many herbivorous mammals. The lip can be folded in through diastema, not allowing the food to fall out.

Diastole: Phase of heartbeat when heart muscles relax and heart refills with blood from veins.

Diastolic pressure: The lowest pressure in an artery during the diastole of heart.

Diathermy: Method of medical treatment of heating the body tissues by the passage of high frequency electric discharge.

Dicentric Of a chromosome or chromatid having two centromeres,

Dichasium: A cyme, with each branch giving rise to two other branches.

Dichlamydeous: When a flower has calyx and corolla.

Dichogamy: The maturing of the anthers and ovules in the same flower at different times, to ensure that self-pollination does not occur.

Dichotomous branching: Forked branching, produced by the subdivision of an apical meristem, to form two branches of the same size.

Diclinous (unisexual): Male and female organs borne on different plants.

Dicotyledoneae (Dicotyledons): One of the two classes of the Angiosperms. The embryo has two cotyledons; the leaves have reticulate venation; the stems have open bundles; and the flower parts are usually in fours or fives.

Dictyoptera: An order of insects comprising the cockroaches and the mantids.

Dictyosome: These are the several interconnected subunits of Golgi Complex in plant cells. Help in synthesis of cell wall and in pushing wastes out of the cell.

Dictyostele: Amphiphloic siphonostele that is broken by crowded leaf gaps into a network of distinct vascular strands (meristeles), each surrounded by an endodermis. Present in stems of certain ferns.

Didymous: (1) In pairs. (2) Of a fruit composed of two similar parts slightly attached along the edge.

Didynamous: Four stamens in two pairs. Those of one pair longer than those of the other, e.g. family Labiatae.

Diencephlon: Hind portion of the vertebrate forebrain. It comprise,,, the optic tracts, thalami, pineal and pituitary.

Diestrus: Short period of quiescence immediately following estrus in the mammalian sexual cycle.

Differentiation: Process of change in cells, tissues or organs during development (embryonic, regenerative or other) resulting in the appearance, where they were previously lacking, of the structures and functions that characterize the kinds of cells, tissues or organs i . different parts of an adult, change in the reverse direction is dedifferentiation.

Diffuse: (1) A prostrate stem that branches freely and loosely, spreading over a wide area. (2) When parenchymatous cells are scattered throughout the xylem.

Diffuse growth: The growth of an algal thallus by the division of any of its cells.

Diffusion: Movement of molecules from the region of their higher concentration to the region of their lower concentration until the concentration is equalised.

Diffusion pressure deficit. (D.P.D.) The osmotic pressure of a solution particularly when comparing it with that of another solution; e.g. a comparison of the osmotic pressure of the cell-sap compared with that of the surrounding medium.

Digenic: Of hereditary difference determined by two genes.

Digestion: Breakdown of complex food by enzymes to simpler compounds which can be incorporated into metabolism.

Digestive system: The system which receives food and prepares it for absorption.

Digitigrade: Walking on toes, not on whole foot, e.g. cat, dog.

Digitoxin: An alkaloid obtained from *Digitalis,* used in the treatment of certain heart conditions.

Dihybrid cross: Cross in which inheritance of two pairs of contrasting characters is studied at a time.

Dikaryon: A fungal hypha whose cells contain two haploid nuclei that divide simultaneously.

Dikaryophase: A diploid phase having a dikaryon.

Dimegaly: Possessing sperms of two sizes.

Dimer: A polymer of two monomer units.

Dimorphism: Difference of form between two members of a species, as between males and females.

Dinosaurs: Extinct reptiles of the orders Saurischia and Ornithischia of the subclass Archosauria, evolved from thecodonts dominant terrestrial in Jurassic and Cretaceous; some were bipedal (*Anatosaurus*) while others were quadrupedal (*Torosaurus* and *Barosaurus*).

Dioecious: With the male and female organs in separate individuals.

Diopter: The unit of refractive power of lens of the eye.

Diptheria: An acute contagious disease caused by the diptheria bacillus which gain entry through respiratory passage.

Diphygenic: Having two types of development.

Diphyletic: Descending from two distinct ancestral groups.

Diplanetism: (1) Occurrence of two zoospore stages, whether morphologically: distinct or identical, separated by encystment. (2) Applied only to former condition. Characteristic of asexual reproduction of some Phycomycete fungi.

Dipleurula: A bilaterally symmetrical larva of echinoderms. Auricularia, bipinnaria and pluteus are forms of dipleurula larvae.

Diplobiont: A plant which has two different kinds of individuals in its life-cycle. If the species is dioecious, there will be three kinds.

Diploblastic: Animals having only two germinal layers during development i.e. ectoderm and endoderm. Therefore, all body organs are derived from these two germinal layers as in Porifera, Coelenterata and Ctenophora.

Diplodization (Diploidization): The conversion of a mycelium containing uninucleate cells into one containing binucleate ones.

Diplohaplont: Having a life-cycle in which a many-celled haploid general ion alternates with a similar diploid one.

Diploid: Chromosome number double to that of a gamete of a given species. Chromosomes are present in pairs.

Diploid apogamy (Euapogamy): The development of a sporophyte containing diploid nuclei from one or more cells of the gametophyte, without any preliminary fusion of gametes,

Diplonema: (1) The stage in the meiotic division at which the chromosomes are clearly double. (2) A small diploid plantlet in the life-cycle of some of the Phaeophyta.

Diplont: (1) An organism at the diploid stage of the lifecycle of plants. (2) The stage in the lifecycle of the Basidiomycetes where the cells have two nuclei

Diplopia: Double vision; seeing one object as two.

Diplopoda: A class of Arthropoda, including millipedes, with wormlike animals hying a head and a segmented body with 2 pairs of legs per segment, head with a pair of antennae and sometimes a pair of eyes.

Diplosome: A paired heterochromosome; a double centrosome lying in the cytoplasm, a pair of centrioles.

Diplotene: Stage in prophase of first division of meiosis, following pachytene, in which pairs of chromatids derived from homologous chromosomes begin to separate from each other except at certain points of connection (chiasmata) where interchange occurs between chromatid segment.

Diptera: Two-winged flies; a large and varied order of holometabolous insects with a single pair of membranous wings and a reduced hind pair modified to form halteres or balancers. The mouth parts are adapted for sucking blood. Many transmit diseases to animals and humans either by bloodsucking or by carrying the germs on their bodies.

Disaccharide: A sugar composed of two monosaccharides. The biologically important disaccharides have 12 carbon atoms formed from two hexoses, e.g. sucrose, maltose, lactose.

Discus proligerus: The oocyte of the mammal is attached to the inner wall of the follicle by its neck which along with the cells surrounding the ovum, are together known as the discus proligerus.

Disinfect To free from infection by destruction of the pest or pathogen.

Disjunction: Separation of chromosomes at anaphase.

Disjunctor: A small cell between two neighbouring conidia in a chain. It disintegrates, thus aiding in dispersion.

Dislocation: The displacement of any part from the normal position especially with reference to joints.

Dispersal: Movement of individuals out of a population or into a population.

Dissimilation: The chemical disintegration of protoplasm, usually by oxidation, with release of energy; catabolism.

Dissociation: Separation of ions from a molecule of crystal lattice.

Distal: Farthest from the point of origin. Part of a chromosome arm, which is farther from the centromere than another part

Diuretic: A compound which increases the activity of the kidneys.

Divergent: Separating from a common source.

Diverticulum: Blind-ending, tubular or saclike out-pushing from a cavity.

DNA: A nucleic acid, mainly found in the chromosomes, that contains the hereditary information of organisms. The molecule is made up of twohelical polynucleotide chains coiled around each other to give a double helix.

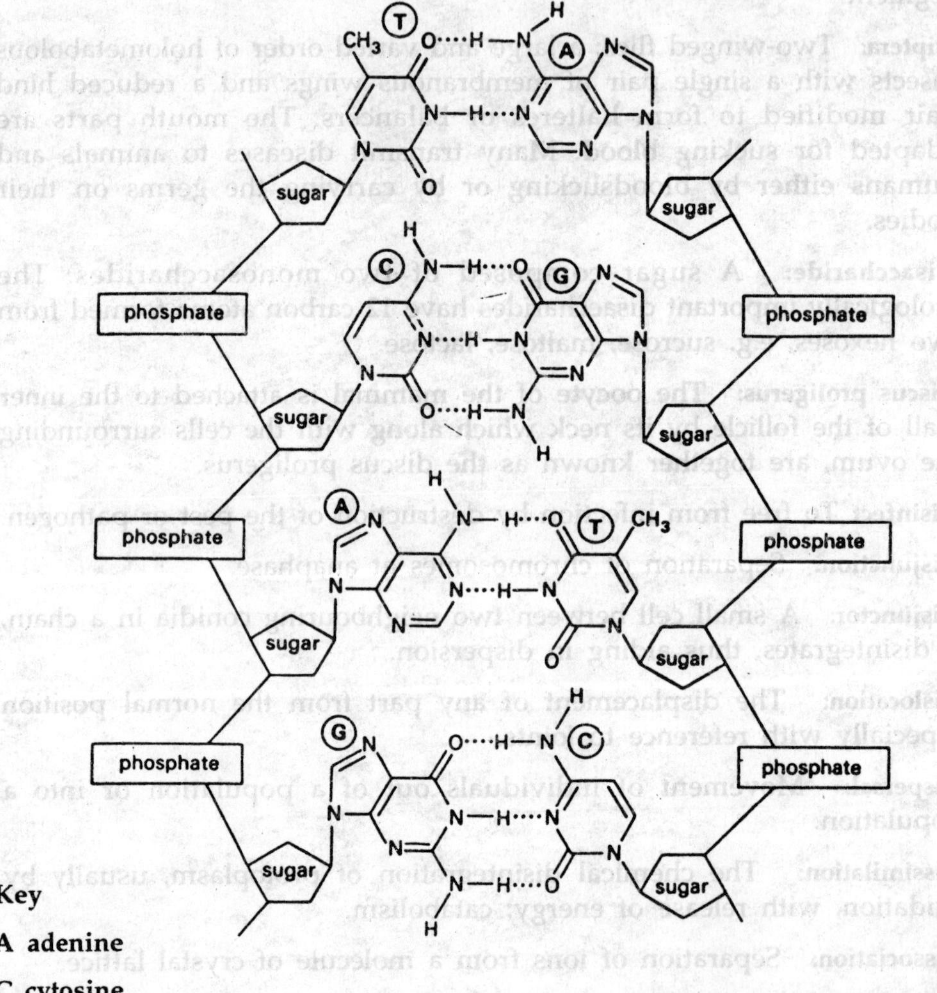

Part of the structure of DNA

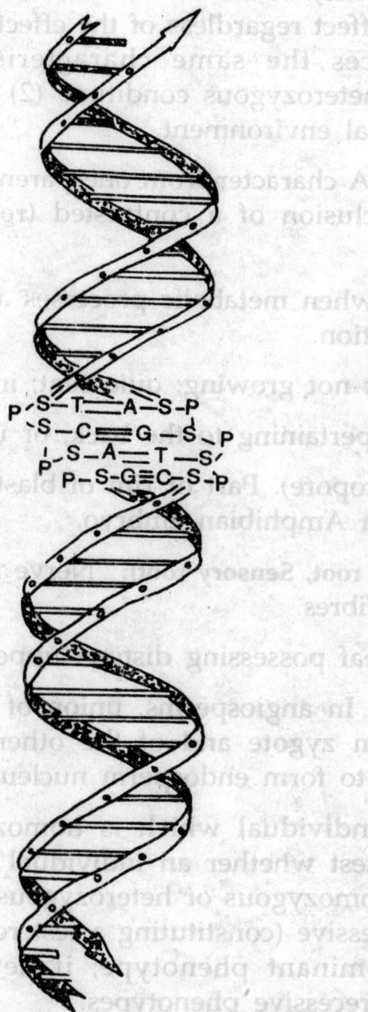

S-P sugar-phospahte chain═ hydrogen bonds linking bases

Domestication: Process of bringing about desirable genetic changes in wild plants and animals for yield, quality, hardiness, etc., through continuous breeding and selection over hundreds of years.

Dominance: (1) The relationship of two allelomorphs where the single gene heterozygote resembles one of the two homozygous parents in phenotype, rather than the other. (2) The prevalence of one or a few species in a community, influencing its general appearance, and the other plants in the community.

Dominant: (1) (Genetics). A functional attribute of a gene. A dominant gene exerts its full effect regardless of the effect of its allelic (recessive) partner. It produces the same characteristic when present in homozygous or in heterozygous condition. (2) (Plant Ecology) Species that controls the local environment.

Dominant character: A character from one parent that manifests itself in offspring to the exclusion of a contrasted (recessive) character from other parent.

Dormance: A state when metabolic processes are slowed. This applies especially to respiration.

Dormant: Alive, but not growing; quiescent; inactive; a resting stage.

Dorsal: Toward of pertaining to the back, or upper surface.

Dorsal lip: (Of blastopore). Part of rim of blastopore corresponding to future dorsal side in Amphibian embryo.

Dorsal root (Posterior root, Sensory root): Nerve root of vertebrates containing the sensory fibres.

Dorsiventral leaf: Leaf possessing distinct upper and lower surface.

Double fertilization: In angiosperms, union of one male nucleus with egg nucleus to form zygote and of the other nucleus with primary endosperm nucleus to form endosperm nucleus.

Double recessive: Individual which is homozygous for a particular recessive gene. To test whether an individual showing character of a dominant gene is homozygous or heterozygous, it is crossed to corresponding double recessive (constituting a testcross); if homo-zygous, all offspring show dominant phenotype; if heterozygous, half show dominant and half recessive phenotypes.

Down feather Nestling downs): A type of feathers in birds, covering the newly hatched bird; found beneath the contour feathers.

Down's syndrome: It is a congenital defect caused by trisomy of 21st chromosome pair (an extra chromosome): It is characterised by flattened skull, protruding tongue and mental retardation. It is also known as Mongolism.

Drift: Changes in the aggregate of genotypes in a small population resulting from the random extinction of allelomorphs of genes in regard to which the population was heterozygous.

E

Ear: The sense organ in vertebrates that is specialized for the detection of sound and the maintenance of balance. It can be divided into the outer ear and middle ear, which collect and transmit sound waves, and the inner car, which contains the organs of balance and (except in fish) hearing.

Ear ossicles: Three small bones—the *incus (anvil), malleus (hammer)*, and stapes (stirrup)—that lie in the mammalian middle ear, forming a bridge between the tympanum (eardrum) and the fenestra ovalis. The function of the ossicles is to transmit (and amplify) vibrations of the tympanum across the middle ear to the fenestra ovalis, which transfers them to the inner ear.

Ecdysis (moulting): The periodic loss of the outer cuticle of arthropods. It stub with the reabsorption of some materials in the inner part of the old cuticle and the fornation of a new soft cuticle. The remains of the old cuticle then split; the animal emerges and absorbs water or swallows air and increases in size while the new cuticle is still soft. This cuticle is then hardened with chitin and lime salts. In insects and crustaceans eedysis is controlled by the hormone eodysone. 2. The periodic shedding of the outer layer of the epidermis of reptiles to allow growth to occur.

Ectlysone: A steroid hormone, produced by insects and crustaceans, that stimulates moulting and metamorphosis, It acts on specific gene loci, stimulating the synthesis of proteins involved in these bodily changes.

Echinodennata: A phylum of marine invertebrates that includes the sea urchins, starfish, brittlestars, and sea cucumbers. Echinoderms have an exoskeleto,i (test) of calcareous plates embedded in the skin. In many species spines protrude from the test. A system of waterfilled canals:

provides hydraulic power for thousands of *tube feet:* sachke protrusions of the body wall used for locomotion, feeding. and respiration. Echinoderms have a long history.

Echolocation: A method used by some animals (such as bats, dolphins, and certain birds) to detect objects in the dark. The animal emits a series of highpitched sounds that echo back from the object and are detected by the car or some other sensory receptor. From the direction of the echo and from the time between emission and reception of the sounds the object is located, often very accurately.

Ecological niche: The status or role of an organism in its environment. An organism's niche is defined by the types of food it consumes, its predators, temperature tolerances,. etc. Two species cannot coexist stably if they occupy identical hiches.

Ecology: The study of the interrelationships between organisms and their natural environment, both flying and nonliving. For this purpose, ecologists study organisms in the context of the populations and communities in which they can be grouped and the ecosystems of which they form a part, The study of ecological interactions .provi-des important information on the nature and mechanisms of evolutionary change.

Ecosystem: A biological community and the physical environment associated with it. Nutrients pass between the different organisms in an ecosystem in definite pathways; for example, nutrients in the soil are taken up by plants, which are then eaten by herbivores, which in turn may be eaten by carnivores: Organisms are classified on the basis of their position in an ecosystem into various trophic levels. Nutrients and energy move round ecosystem in loops or cycles (in the case above, for example, nutrients are returned to the soil via animal wastes and decomposition).

Ectoderm: The external layer of cells of the gastrula, which wig develop into the epidermis and the nervous system in the adult.

Ectoparasite: A parasite that fives on the outside of its host's body.

Edaphic factor: A factor relating to the physical or chemical composition of the soil found in a particular area. For example, very alkaline soil may be an edaphic factor limiting the variety of plants growing in a region.

Effector: A cell or organ that produces a Physiological response when stimulated by a nerve impulse. Examples include muscles and glands.

Efferent: Carrying (nerve impulses, blood, etc.) away from the centre of a body or organ towards peripheral regions, The term is usually applied to types of nerve fibres or blood vessels.

Eftmion: The flow of a gas through a small aperture. The relative rates at which gases effuse, under the same conditions, is approximately inversely proportional to the square roots of their densities.

Egg 1: The fertilized ovum (.zygote) in Mlaying animals, e.g. birds and insects, after it emerges from the body. The egg is covered by egg membranes that protect it from environmental damage, such as drying. 2. (or egg cell) The mature female reproductive cell in animals and plants.

Egg membrane: The layer of material that covers an animal egg cell. *Primary membranes* develop in the ovary and cover the egg surface in addition to the normal cell membrane. The primary membrane is called the *vitelline membrane* in insects, molluses, birds, and amphi-bians, the *chorion* in tunicates and fish, and the *zona pellucida* in mammals. Insects have a second thicker membrane, also called the chorion, *Secondaly membranes* are secreted by the oviducts and parts of the genital system while the egg is passing to the outside.

Elasmobranchli (Chondrich-thyes): A class of vertebrates comprising the fishes with cartilagmous skeletons. The majority belong to the order Selachi (skates, rays, and sharks). Most cartilagmous fishes are marine carnivores with powerful jaws. Unlike bony fishes, they have no swim bladder, and therefore, avoid sinking only by constant swimfrung with the aid of an asymmetrical *(heterocercal)* tail. There is no operculum covering the gill slits, the, first of which is modified as a spiracle internal so the few eggs Produced are consequently yolky, large, and well-protected. Some elasmobranchs show viviparow development of the young.

Electric organ: An organ occurring on the body or tail of certain fish, such as the electric ray *(Torpedo)* and electric eel *(Electrophorus elertricus).* It gives an electric shock when touched and is used either to stun prey or predators or, in some species, to maintain a weak electric field in the surrounding water that is used in navigation. 1he organ is composed of modified muscle cells *(electroplate cells),* nervous stimulation of which greatly increases the potential difference across the cell.

Electrocardiogram (ECG): A tracing or graph of the electrical activity of the heart, Recordings are made front electrodes fastened over the heart and usually on both arms and a leg. Changes in the normal pattern of an ECG may indicate heart irregularities or disease.

Electroencephalogram (EEG): A tracing or graph of the electrical activity of the brain. Electrodes taped to the scalp record electrical waves from different parts of the brain. The pattern of an EEG reflects an individual's Cvei of consciousness and can be used to detect such disorders as epilepsy, turnours, or brain damkge.

Electron microscope: A form of microscope that uses a beam of electrons instead of a beam of light (as in the optical rnicroscope) to form a large image of a very small object, such as a cell organelle, a virus, or a DNA molecule. In optical microscopes the resolution is limited by the wavelength of the fight. High-energy electrons, however, can be associated with a considerably shorter wavelength than light; for example, electrons accelerated to an energy of 10^5 electronvolts have a wavelength of 0.04 nanometre, enabling a resolution of 0.2-0.5 run to be achieved. The transmission electron microscope has an electron beam, sharply focused by electron lenses (coils producing a magnetic field or electrodes between which an electric field is created), passing through a very thin metallized specimen (less than. 50 nanometres thick) onto a fluorescent screen, where a visual image is formed. This image can be photograph. The *scanning electron microscope* can be used with thicker specimens and forms a perspective image, although the resolution and magnification are lower. It is used particularly for examine source of electronsing surface features of small objects, such as pollen grains. In this type of instrument a beam of electrons scans the specimen and those that are reflected, together condenser lens with any secondary electrons emitted, are collected. This current is used to modulate a separate electron beam in a TV monitor, which searts the screen at the same frequency, consequently building up a picture of the specimen. The resolution is limited to about 10–20 nm.

Electron transport chain (respiratory chain): A sequence of biochemical reduction-oxidation reactions that forms the final stage of aerobic respiration. It results in the transfer of electrons or hydrogen atoms derived from the Krebs cycle to molecular oxygen, with the formation of water. At the same time it conserves energy from food or light in the form of ATP. The chain comprises a series of electron carriers that undergo reversible reduction oxidation reactions, accepting electrons and then donating them to the next carrier in the chain. In the mitochondria, NADH and FADH, generated by the Krebs cycle, transfer their electrons to a chain compris-ing flavin mononuclecitide (FMN), coenzyme Q, and a series of eytochromes. This process is coupled to the formation of ATP at three sites along the chain.

Electrophoresis (cataphoresis): A technique for the analysis and separation of col, loids, based on the movement of charged colloidal particles in an electne field. There various experimental methods. In one the sattle is placed in a U-tube and a buffer solution added to each am so the them are sharp boundaries between buffer and sample. An electrode is placed in each arm a voltage applied, and the motion of the boundaries under the influence of the field is observed. The rate of migration of the particles depends on the field, the charge on the particles, and on other factors, such as the sin and shape of the particles. More simply, electrophoresis can be carried out using an adsorbent, such as a strip of filter paper, soaked in a buffer with two electrodes making contact. The sample is placed between the electrodes and a voltage applied. Different components of the mixture migrate at different rates, so the sample separates into zones. The components can be identified by the rate at which they move.

Elytra: The thickened homy forewings of the Coleoptera (beetles), which cover and protect the membranous hind wings when the insect is at rest.

Emasculation: The removal of the anthers of a flower in order to permit self-pollination or the undesirable pollination of neighbouring plants.

Embryo: 1. An animal in the ealiest stage of its development, from the time when the fertilized ovum starts to divide, while it is contained within the egg or reproductive organs of the mother, until hatching or birth. A human embryo is called a fetus after the first eight weeks of pregnancy. 2. The structure in bryophytes, pteridophytes, and seed plants that develops from die zygote prior to germination. In seed plants the zygote is situated in the embryo sac of the ovule It divides by mitosis to form the embryonic cell and a suwture called the suspensor, which embeds the embryo in the surrounding nutritive tissue. The embryonic cell divides continuously and eventually gives rise to the radide (young root), plumule (young shoot), and one or two cotyledons (seed leaves). Changes also take place in the surrounding tissues of the ovulc, which becomes the seed enclosing the embryo plant.

Embryology: The study of the development of animals from the fertilized egg to die new adult organism. It is sometimes limited to the period between fertilization of the egg and hatching or buth.

Embryo sac: A large cell that develops in the ovule of flowering plants. It is equivalent to the female garnetophyte of lower plants, although it is very much reduced. Typically, it contains eight 'nuclei formed by division of the original female gamete. One, the sphere (egg nucleus), is fertilized by a male nucleus and becomes the embryo. Two of the remaining nucleus fuse with a second inhale nucleus to form a triploid nucleus that gives rise to the endosperm

Emulsion: A colloid in which small particles of one liquid are dispersed in another liquid. Usually emulsions involve a dispersion of water in an oil or a dispersion of oil in water, and are stabilized by an difter. Commonly emulsifieris are substances, such as detergents, that have lyophobic and Iyophific: parts in their molecules.

Enamel: The material that forms a covering over the crown of a tooth. Enamel is smooth, white, and extremely hard, being rich in minerals containing calcium. It is produced by certain cells of the oral epithelium and protects *the der*lying dentine of the tooth. Enamel may also be found in the placoid scales of certain fish, which demonstrates the common developmental origin of scales and teeth.

Encephalin (enkephalin): A peptide neurotransnutter found principally in the brain, Two encephalins have so far been isolated. They bind to opiate receptors in the brain and their release is thought to control levels of pain and other sensations. Encephalitis have also been found in spinal cord nerve cells and in gut epithelial cells.

Endocrine gland (ductless gland): gland in an animal that manufactures hcrmones and secretes them directly into the bloodstream to act at distant sites in the body. Endoctrine glands tend to control slow long-term activities in the body, such as growth and sexual development. In mammals they include the pituitary, adrenal, thyroid, and parathyroid glands, the ovary and testis, the placenta, and part of the pancreas. The activity of endocrine glands is controlled by negative feedback, i.e. a rise in output of hormone inhibits a further increase in its Production, either directly or indirectly via the target organ or cell.

Endocrinology: The study of the structure and functions of the endocrine glands and of the hormones they produce.

Endoderm (entoderm): The internal layer of cells of the gastrula, which will develop into *the* alimentary canal (gut) and digestive glands of the adult.

Endodermis: The innermost layer of the root cortex of a plant, lying immediately the outside the vascular tissue. Various modification of the endodermal: walls indicate that they regulate the passage of materials both and out of the vascular system. An endodermis may also be seen in the stems of some plants.

Eye: An organ of sight, or light perception. Invertebrates usually have eyes that are simple photoreceptors, sensitive to the direction and intensity of light.

Eyespot: A light-sensitive structure of some unicellular and colonialalgae and their gametes and zoospores. It contains globules of orange or red carotenoid pigments. It controls lovomotion, ensuring optimum light conditions for photosynthesis. Its location varies. In Chlamydomonas it is just inside the chloroplast, in Euglena it is near the base of the flagellum. A light-sensitive pigmented spot found in the cells of some primitive animals,including Protozoa, jellyfish, and flatworms, e.g., the miracidium larva of liver fluke.

F

F_2 (second filial generation): The sword generation of offspring in breeding experiments, obtained by crosses between individuals of the: generation.

Facilitated diffusion: The transport of molecules across the outer membrane of a living cell, by a process that involves a specific carrier located within the cell membrane but does not require expenditure of energy by the cell. The carrier is believed to combine with a molecule on one side or the membrane, move through the membrane and re lease the molecule on the other side.

FAD (flavin adenine dinucleotide): A coenzyme important in various biochemical reactions. It comprises a phosphorylated vitamin B_2 (riboflavin) molecule linked to the nucleotide adenine monophosphate (AMP). FAD is usually tightly bound to the enzyme forming a *flavoprotein*. It functions as a hydrogen acceptor in dehydrogenations, being reduced to $FADH_2$. This in turn is oxidized to FAD by the electron transport chain, by generating ATP (two molecules of ATP per molecule of $FADH_2$).

Faces: Waste material that is eliminated from the alimentary canal through the anus. Faeces consist of the indigestible residue of food that remains after the processes of digestion and absorption of nutrients and water have taken place, together with bacteria and dead cells shed from the gut bring.

Fallopian tube (oviduct): The tube that carries egg cells from the ovary to the womb in mammals. The eggs are carried by the action of muscles and cilia. It was named after the Italian anatomist Gabriel Fallopius.

Family: A category used in the classification of organisms that consists of one or several similar or closely related genera. Similar families are grouped into an order. Family names end in *-aceae* or *-ae* in botany (e.g. Cactaceae) and *-idae* in zoology (e.g. Equidae).

Fascia: A sheet of fibrous connective tissue occurring beneath the skin and also enveloping glands, vessels, nerves, and forming tendon sheaths.

Fast green: A green dye used in optical microscopy that stains cellulose, cytoplasm, collagen, and mucus green. It is frequently used to stain plant tissues, with safranin as a counterstain, Unlike light green, a similar dye, it does not fade easily.

Fat: A mixture of lipids, chiefly triglycerides, that is solid at normal body temperatures. Fats occur widely in plants and animals as a means of storing food energy, having twice the calorific value of carbohydrates. In mammals, fat is deposited in a layer beneath the skin (subcutaneous fat) and deep within the body as a specialized adipose tissue. The insulating Properties of fat are also important, especially in animals lacking fur and those inhabiting cold climates.

Fats derived from plants and fish generally have a greater proportion of unsaturated fatty acids than those from mammals, Their melting points thus tend to he lower, causing a softer consistency at room temperatures. Highly unsaturated fats are liquid at room temperatures and are therefore more properly called oils.

Fat body: 1. An abdominal organ in amphibians attached to the anterior of each kidney. It contains a reserve of fat that nourishes the gonads during the hibernation in readiness for the spring breeding season. 2. A mass of fatty tissue spreading throughout the body cavity of insects in which fats, proteins, and glycogen are stored as a reserve for hibernation or pupation.

Fatty acid: An organic compound consisting of a hydrocarbon chain and a terminal carboxyl group. Chain length ranges from one hydrogen atom (formic acid, HCOOH) to nearly 30 carbon atoms. Acetic, propionic, and butyric acids are important in metabolism. Long-chain fatty acids (more than 8-10 carbon atoms) most commonly occur as constituents of certain lipids, notably glycerides, Phospholipids, sterols, and waxes, in which they are esterified with alcohols. These long-chain fatty acids generally have an even number of carbon atoms; unbranched predominate over branched chains. They may be saturated (e.g. Palmitic acid and stearic acid) or unsaturated, with one double bond (e.g. oleic acid) or two or more double bonds, in which case they are called *polyunsaturated fatty acids* (e.g. linoleic acid and linolenic acid).

Feathers: The body covering of birds, formed as outgrowths of the epidermis and composed of the protein keratin. Feathers provide beat insulation, they give the body its streamlined shape, and those of the wings and tail are important in flight Basically a feather consists of a *quill,* which is embedded in the skin attached to a feather follicle and is continuous with the shaft (*rachis*) of the feather, which carries the barbs. This basic structure is modified depending on the type of feather.

Fecundity: The fertility of an organism (in higher animals, generally of the female of the species). Normally all organisms, assuming they reach reproductive age, are sufficiently fecund to replace themselves several times over. Darwin noted this, together with the fact that population numbers nevertheless tended to remain fairly constant: these observations led him to formulate his theory of evolution by natural selection.

Feedback: The use of part of the output of a system to control its performance. In positive *feedback,* the output is used to enhance the input; in *negative feedback,* the output is used to reduce the input. Many biological processes rely on negative feedback. As the population of a species expands, so its food supply pet individual is diminished; the result is that the population then begins to fall. Many biochemical processes are controlled by feedback inhibition.

Fehling's test: A chemical test to detect reducing sugars and aldehydes in solution, devised by the German chemist H.C. von Felhing. Fehling's solution consists of Felhings A (copper(II) sulphate solution) and Felhing's B (alkaline 2,3dihydroxybutanedioate (sodium tartrate) solution), equal amounts of which are added to the test solution. After boiling, a positive result is indicated by the formation of a brick-red precipitate of copper(l) oxide.

Femur 1. The thigh bone of terrestrial vertebrates. It articulates at one end with the pelvic at the hip joint and at the other (via two condyles) with the tibia.

Fenestra: Either of the two delicate membranes between the middle ear and the Inner ear. The lower membrane is the *fenestra ovalis* (fenestra *vestibuli, or oval window).* The stapes (the third car ossicle) transmits to the fenestra ovalis the vibrations it has received (via the other two car ossicles) from the tympanum (eardrum). Vibrations in the fenestra ovalis are transmitted to the cochlea in the inner ear.

Fermentation: A form of anaerobic respiration occurring in certain microorganisms, e.g. yeasts. It comprises a series of biochemical reactions by which pyruvate (the end product of glycolysis) is converted to ethanol and carbon dioxide.

Fertilization (syngamy): The fusion of the nuclei of the male and female gametes (productive cells) during the process of sexual reproduction to form a *zygote*. As each gamete contains only half the correct number of chromosomes, fertilization and zygote formation results in a cell with the full complement of chromosomes, half of which are derived from each of the parents. In as the process involves fusion of the nuclei of a spermatozoan and an ovum. In most aquatic animals (e.g. fish) this takes place in the surrounding water, into which the gametes are shed. Among most terrestrial animals (e.g. insects, many mammals) fertilization occurs m the body of the female, into which the sperms are introduced. In flowering plants, after pollination, the male gamete (pollen) produces a pollen tube, which grows down into the female reproductive organ (carpel) to enable a pollen nucleus to fuse with the egg nucleus.

Fetus (foetus): The embryo of a mammal, especially a human, when development has reached a stage at which the main features of the adult form are recognizable. In humans the embryo from eight weeks to birth is called a fetus.

Feulgen's test: A histochemical test in which the distribution of DNA in the chromosomes of dividing cell nuclei can be observed. It was devised by the German chemist R. Feulgen. A tissue section is first treated with dilute hydrochloric acid to remove the purine bases of the DNA, thus exposing the aldehyde groups of the sugar deoxyribose.

Fibre: An elongated plant cell whose walls are extensively (usually completely) thickened with lignin. Fibres are found in the vascular tissue, usually in the xylem, where they provide structural support. The term is often used loosely to mean any kind of xylem element. The fibres of many species, e.g. flax, are of commercial importance.

Fibrin: The insoluble protein that forms fibres at the site of an injury and is the foundation of a blood clot.

Fibrinolysis: The protein dissolved in the blood plasma that, when suitably activated, is converted to insoluble fibrin fibres. See clotting factors.

Fibribrinolysis: The breakdown of the protein fibrin by the enzyme *plasmin* (or *fibrinolysin),* which occurs when blood clots are removed from the circulation.

Fibroblast: A cell that secretes fibres in the intercellular substance of connective tissue. The cells are long, flat, and star-shaped and lie close to collagen fibres.

Fibula: The smaller and outer of the two bones between the knee and the ankle in terrestrial vertebrates.

Field-emission microscope: A type of electron microscope in which a high negative voltage is applied to a metal tip placed in an evacuated vessel some distance from a glass screen with a fluorescent costing. The tip produces electrons by *field emission,* i.e. the emulsion of electrons from an unheated sharp metal part as a result of a high electric field. The omitted electrons form an enlarged pattern on the fluorescent screen, related to the individual exposed planes of atoms. As the resolution of the instrument is limited by the vibrations of the metal atoms, it is helpful to cool the tip in liquid helium.

Field-ionization microscope (field-ion microscope): A type of electron microscope that is similar in principle to the field-emission microscope, except that a high positive voltage is applied to the metal tip, which is surrounded by low-pressure gas (usually helium) rather than a vacuum. The image is formed in this case by *field ionization*: ionization at the surface of an unheated solid as a result of a strong electric field creating positive ions by electron transfer from surrounding atoms or molecules. The image is formed by ions striking the fluorescent screen.

Filament: (in zoology) A long slender hairlike structure, such as any of the barbs of a bird's feather. 2. (in botany) The stalk of the stamen in a flower. It bears the anther and consists mainly of conducting tissue.

Filicinae: A class of mainly terrestrial vascular plants —the ferns— belonging to the Pteropsida. Ferns are perennial plants bearing large conspicuous leaves (fronds) usually arising from either a rhizome or a short erect stern. Bracken is a common example. Only the tree ferns have stem that reach an appreciable height. There is a characteristic uncurling of the young leaves as they expand into the adult form. Spores are borne on the underside of specialized leaves (sporophylls).

Filoplumes: Minute hairlike feathers consisting of a shaft (rachis) bearing a few unattached barbs. They are found between the contour feathers.

Filter feeding: A method of feeding m which tiny food particles are ingested from the surrounding water. It is used by many aquatic invertebrates.

Filtration: The process of separating solid particles using a filter. In vacuum filtration, the liquid is drawn through the filter by a vacuum pump.

Fins: The locomotory organs of aquatic vertebrates. In fish them are typically one or more dorsal and ventral fins (sometimes continuous), whose function is balance; a caudal *fin* around the tail, which is the main propulsive organ; and two paired fins: the pectoral fins attached to the pectoral (shoulder) girdle and the *pelvic fins* attached to the pelvic (hip) girdle, which are used in steering. These paired fins are homologous with the limbs of tetrapods.

Fission: A type of asexual reproduction ocurring in some unicellular organisms, e.g., diatoms, protozoans, and bacteria in which the parent cell divides to form two (binary fission) or more (*multiple fission*) similar daughter cells.

Fission-track dating: A method of estimating the age of glass and other mineral objects by observing the tracks made in them by the fission fragments of the uranium nuclei that they contain By irradiating the objects with neutrons to induce fission and comparing the density and number of the tracks before and after irradiation it is possible to estimate the time that has elapsed since the object solidified.

Fitness: (in evolution) The condition of an organism that is well adapted to its environment, as measured by its ability to reproduce itself.

Fixation 1. The first stage in the preparation of a specimen for microscopical examination, in which the tissue is killed and preserved in as natural a state as possible by immersion in a chemical *fixative*. The fixative prevents the distortion of cell components by denaturing its constituent protein. Some commonly used fixatives are formaldehyde, ethanol, and Bouin's fluid, and osmium tetroxide and gluteraldehyde.

Flaccid: (in botany) Describing plant tissue that has become soft and less rigid than normal because the cytoplasm within its cells has shrunk and contracted away from the cell walls through loss of water.

Flagellum: A relatively long (up to 150 mm) fine whiplike structure present on the surface of certain cells, particularly motile reproductive cells (e.g. spermatozoa), bacteria, and certain protozoans. Flagella occur singly or in small groups. Beating of flagella produces movement of the cell, but some flagella (e.g. in Hydra) beat to cause movement of the surrounding fluid.

Flavonoid: One of a group of naturally occurring phenolic compounds many of which are plant pigments. They include the anthocyanins, flavonols, and flavones. Patterns of flavonoid distribution have been used in taxonomic studies of plant species.

Flower: The structure in angiosperms (flowering plants) that bears the organs for sexual reproduction. Flowers are very variable in form, ranging from the small green insignificant wind-pollinated flowers of many grasses to spectacular brightly: coloured insect-pollinated flowers. Flowers are often grouped together into inflorescences, some of which (e.g. that of dandelion) are so compacted as to resemble a single flower. Typically flowers consist of a receptacle that bears sepals, petals, stamens, and carpels. The flower parts are adapted to bring about pollination and fertilization resulting in the formation of seeds and fruits. The sepals are usually green and leaflike and protect the flower bud. The petals of insect-pollinated flowers are adapted in many ingenious ways to attract insects and, in some instances, other animals. Flowers adapted for pollination by long-tongued insects have a long corolla tube of fused petals with nectar in a concealed position. The tongue of the insect brushes against the anthers and stigma before reaching the nectar. Wind-pollinated flowers, in contrast, are inconspicuous.

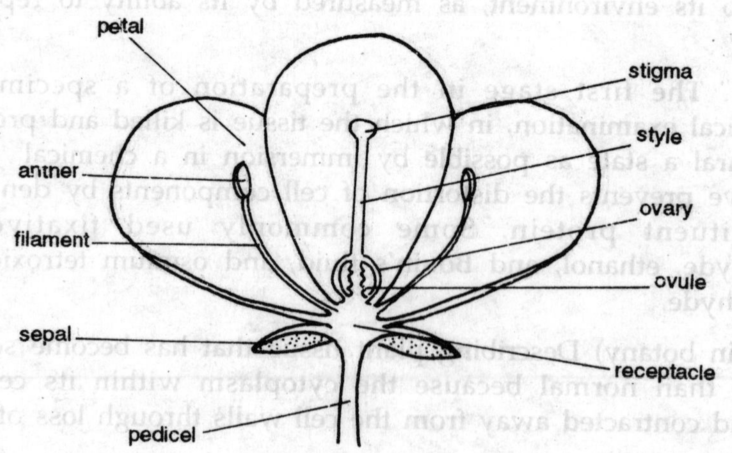

Typical half-flower

Fluordation: The process of adding very small amounts of fluorine salts (e.g. sodium fluoride, NaF) to drinking water to prevent tooth decay. The fluoride becomes incorporated into the substance of the growing teeth and reduces the incidence of caries.

Folic acid (folacin): A vitamin of the vitamin B complex. In its active form, tetrahydrofolic acid, it is a coenzyme in various reactions involved in the metabolism of amino acids, purines, and pyrimidines. It is synthesized by intestinal bacteria and is widespread in food, especially green leafy vegetables.

Follicle 1: (in animal anatomy) Any enclosing cluster of cells that protects and nourishes a cell or structure within. For example, follicles in the ovary surround the developing egg, while hair follicles envelop the roots of hairs. 2. (in botany) A dry fruit that, when ripe, splits along one side to release its seeds. It is formed from a single carpel containing one or more seeds. Follicles do not occur singly but are grouped to form clusters *(etaerios).*

Follicle-stimulating hormone (FSH): A hormone, secreted by the anterior pituitary gland in mammals, that stimulates, in female mammals, ripening of specialized structures in the ovary (Graafian follicles) that produce ova and, in males, the formation of sperm in the testis. It is a major constituent of fertility drugs, used to treat failure of ovulation and decreased sperm production.

Food chain: The transfer of energy from green plants (the ph producers) through a sequence of organisms in which each eats the one below it in the chain and is eaten by the one above. Thus plants are eaten by herbivores, which are then eaten by carnivores. These may in turn be eaten by different carnivores. The position an organism occupies m a food chain is known as its trophic level. In practice, many animals feed at several different trophic levels, resulting in a more complex set of feeding relationships known as a *food web.*

Foramen: An aperture in an animal part or organ, especially one in a bone or cartilage. For example, the *foramen magnum* is the opening at the base of the skull through which the spinal cord passes.

Forebrain: (prosencephalon) One of the three sections of the brain of a vertebrate embryo. The forebrain develops to form the cerebrum, hypothalamus, and thalamus in the adult.

Foregut: 1. The anterior region of the alimentary canal of vertebrates, up to the anterior part of the duodenum. 2. The anterior part of the alimentary canal of arthropods.

Form 1. A category used in the classification of organisms into which different types of a variety may be placed. 2. Any distinct variant within a species. Seasonal variants e.g. the tawny brown (summer) and blue-white (winter) forms, of the blue hare, may be called forms, as may the different types that constitute a polymorphism.

Fossil: The remains or traces of any organism that lived in the geological past. In genera] only the hard parts of organisms become fossilized (e.g. bones, teeth, shells, and wood) but under certain circumstances the entire organism is preserved. For example, virtually unaltered fossils of extinct mammals, such as the woolly mammoth and woolly rhinoceros, have been found preserved in ice in the Arctic.

In the majority of fossils the organism has been turned to stone a process known as *petrification. This* may take one of three forms. In *permineralization,* solutions originating underground fill the microscopic cavities in the organism. Minerals in these solutions (e.g. silica or calcite) may actually replace the original material of the organism so that even microscopic structures may be preserved; this process is known as *replacement* (or *mineralization).* A third form of petrification carbonization (or *distillation)* occurs in certain soft tissues that are composed chiefly of compounds of carbon, hydrogen, and oxygen (e.g., cellulose). After the organism has been buried, and in the absence of oxygen, carbon dioxide and water are liberated until only free carbon remains. This forms a black carbon film in the rock outlining the original organism. Moulds are formed when the original fossil is dissolved away leaving a mould of its outline in the solid rock. The deposition of mineral matter from underground solutions in a mould forms a cast. Paleontologists often produce casts from moulds using such substances as dental wax. Moulds of thin organisms (e.g. leaves) are commonly known as imprints.

The ideal conditions for the formation of fossils occur in areas of rapid sedimentation, especially tow parts of the seabed that lie below the zone of wave disturbance. See also chemical fossil; index fossil; microfossil.

Fovea (fovea centralis): A shallow depression in the retina of the eye, opposite the lens, that is present in some vertebrates. This area contains a large concentration of cones with only a thin layer of overlying nerves. It is therefore specialized for the perception of colour and sharp intense images. The clarity is enhanced when light is focused on the fovea of both retinas simultaneously.

Fragmentation: A method of asexual reproduction, occurring in some invertebrate animals, in which parts of the organism break off and subsequently differentiate and develop into new individuals. It occurs especially in certain coelenterates and annelids. In some, regeneration may occur before separation, producing chains of individuals budding from the parent,

Fraternal twins (dizygotic twins): Two individuals that result from a single pregnancy, each having developed from a separate fertilized egg. The two egg cells contain different combinations of alleles and so do the two sperm that fertilize them. The twins therefore have no more genetic similarity than brothers or sisters from single births.

Fruit: The structure formed from the ovary of a flower, usually after the ovules have been fertilized. It consists of the *fruit wall* enclosing the seed(s). Other parts of the flower, such as the receptacle, may develop and contribute to the structure, resulting in a *false fruit* . The fruit may retain the seeds and be dispersed whole (an *indehiscent fruit),* or it may open (*dehise*) to release the seeds (*a dehiscent fruit).* Fruits are divided into two main groups depending on whether the ovary wall remains dry or becomes fleshy *(succulent).* Succulent fruits are generally dispersed by animals and dry fruits by wind, water, or by some mechanical means.

Fucoxanthin: The major carotenoid pigment present, with chlorophyll, in the brown algae.

Fungi: A group of Simple Plants lacking chlorophyll. Some authorities do not regard fungi as plants and include them with other organisms of uncertain affinities in the kingdom Protista. Fungi are classified in, the taxonomic division Mycota or Fungi. They can either exist as single cells or make up a multicellular body, called a *mycelium,* which consists of filaments known as hyphae, Most fungal cells are multinucleate and have cell walis composed chiefly of chitin. Fungi exist primarily in damp situations on land and because of the absence of chlorophyll, are either parasites or saprophytes on other organisms.

Fungi Imperfecti (imperfect fungi): A class of fungi in which sexual reproduction is absent. Otherwise they are generally similar to Ascomycetes and are consequently thought to he Ascomycetes that have lost the ability to produce ascospores.

Gall bladder: Small bladder in many vertebrates arising from bile duct near or in liver, storing bile. Contractile walls empty bladder, when here is food, especially fat, in intestine, probably activated to do so by hormones secreted by intestine wall.

Gametangium: Any organ in which gametes are produced. Sex organs especially of the Thallophyta.

Gamete: A haploid cell taking part in sexual reproduction. The two gamete nuclei, and frequently the cytoplasm, fuse to form a diploid zygote to initiate the development of a new individual. Female gamete may develop directly without fusion (parthenogenesis).

Gametocyte: Cell that undergoes meiosis to produce gametes.

Gametogenesis: Process of formation and maturation of gametes in the gonads. Spermatogenesis is the gametogenesis occurring in male gonads i.e. testes to produce sperms while oogenesis occurs in the female gonads i.e. ovaries to produce eggs, controlled by hormones from the pituitary gland, e.g. FSH.

Gametophyte: The haploid or gamete-producing generation of any plant. It reproduces sexually and produces diploid sporophyte generation.

Gamma globulin: Proteins of human blood serum consisting of fibrinogen, albumin and several globulins. Gamma globulin contains greatest concentration of antibodies and therefore is an effective agent in providing passive immunity to certain diseases.

Gamont: An individual in the lifecycle of certain protozoans whose subdivision produces gametes.

Gamopetalous (Sympetalous): Flower with united petals, e.g. primrose.

Ganglion: A group of neuronal cell bodies located outside central nervous system of vertebrates. The nervous system of many invertebrates consists largely of ganglia.

Ganoid scale: Dermal scale typical of primitive Actinopterygii. Diamond-shaped and closely arranged in diagonal rows like tiles in a floor, e.g. *Polypterus* and *Lepsosteus.*

Gastrocoel (Archenteron): Primitive digestive cavity of a metazoan embryo, formed by gastrulation.

Gastrodermis: Lining epithelium (endoderm) of the digestive cavity in coelenterates.

Gastropoda: A class of mollusca including snails and slugs. Well developed head with eyes, tentacles and radula present; foot large and flat; shell when present univalve and coiled; mostly marine, some freshwater or terrestrial.

Gastrovascular: Serving for both digestion and circulation.

Gastrula: Stage of the embryonic development of animals during which ectoderm, endoderm, mesoderm and the archenteron are formed.

Gastrulation: Embryo logical term fog the complex cell movements which occur in almost all animals at the end of the cleavage. The movements carry those cells whose descendants will form the future internal organs from their superficial position in the blastula to their definitive positions inside the embryo.

Gause's principle (Principle of exclusion) No two species occupying same niche can coexist in the same place at the same time.

Geitonogamy: The crosspollination between two flowers on the same plant.

Gel: A colloidal system in which the solid phase is continuous and the liquid phase is dispersed. Jellylike state of protoplasm.

Gelatin: A colourless, odourless, tasteless, jellylike protein obtained from bones, horns and hides.

Gemma: Organ of vegetative reproduction, in mosses and liverworts.

Gemmation: Kind of sexual reproduction by formation of a group of new member of a colony of connected individuals.

Gemmule: In sponges a bud formed internally as a group of cells, which may become free by decay of parent and subsequently form a new individual.

Gene: Functional unit of chromosome and responsible for hereditary trait. These are linearly arranged of chromosomes.

Genecology: Study of genetical composition of plant population relation to their habitat.

Gene flow: Movement of genes, as a result of mating and gene exchange within populations.

Gene frequency: Frequency of occurrence of a given kind of gene in a population in relation to frequency of all its alleles.

Gene pool: Sum total of genes of a particular population.

Generation time The time taken by a unicellular organism between one cell-division and the next.

Generative cell: A cell in the pollen grain of gymnosperms which divides to give a stalk cell and a body cell.

Genetic: Pertaining to, or analogous with, heredity.

Genetic code: Sequence of three nucleotides in mRNA (complimentary to triplet sequence in DNA) that code for specific amino acids. There are sixty four triplets in the code.

Genetic complex: The sum total of the hereditary factors contained in chromosomes and cytoplasm.

Genetic drift: Random change in gene frequency from one generation to another in a population.

Genetic engineering: Artificial alteration of the genetic makeup of cells.

Genetic equilibrium: Condition of a population in which successive generations consist of the same genotype with the same frequencies, in respect of particular genes, or arrangement of genes.

Genetics: Science of heredity, including the study of its chemical foundation, its developmental expression and its bearing on variation, selection, adaptation, evolution etc.

Genic balance: Hypothesis that the characters of an organism are each determined by the interaction of a large, but unknown number of genes, some affecting development in one direction and some in another, so that the ultimate result is a balance struck between the total effect.

Genome: A set of chromosomes inherited from one parent.

Genotype: Genetic constitution of an organism.

Genus: One of the kinds of group used in classifying organisms. Consists of a number of similar species; occasionally of one species only.

Geocarpy: Ripening of fruits underground. The young fruits are pushed into the soil by a post-fertilization curvature of the stalk.

Geological periods: Time chart of animal or plant life. Estimate of the age of fossils are derived mainly from study of radioactive minerals in fossil-bearing rocks. The whole geological column has been divided into five eras, i.e. Archcozoic, Proterozoic, Paleozoic, Mesozoic and Coenozoic: Each era is divided into many periods. Currently it is the Coenozoic era.

Geophyte: A plant which perennates by subterranean buds.

Geotaxis: Taxis in which stimulus is gravitational force.

Geotropism: The response of part of plant due to the stimulus of gravity.

Germ cells: Cells which give rise to male and female gametes.

Germicide: Substance capable of destroying bacteria.

Germinal disc: Small protoplasmic disc on the surface of huge yolk of the hen's egg.

Germinal epithelium: Epithelium lining seminiferous tubules of vertebrate testis, from which during sexual maturity arise the spermatogonia and the sertoli cells; applied also to epithelium covering coelomic surface of vertebrate ovary, which probably forms follicle cells, though it is now doubtful whether it forms oogonia throughout sexual maturity.

Germinal variations: Variations due to some modification in the germ cells.

Germ layer: One of the three primary layers which can be distinguished in an embryo during and immediately after gastrulation. These are two in diploblastic animals, endoderm and ectoderm. In triploblastic animals mesoderm is also present.

Germplasm: A particular sort of protoplasm which Weismann suggested was transmitted unchanged from generation to generation via the germ cells, giving rise in each individual to the body cells (soma) but itself remaining distinct and unaffected by the environment of the individual.

Gerontology: Study of bid age of man.

Gestation period: Length of time from conception to birth in viviparous animals.

Giantfibre: Nerve fibre of very large diameter (upto 1 mm in squids) occurring in many invertebrates (e.g. annelids, crustaceans, cephalopods) and some vertebrates, providing very rapid conduction of impulses for sudden locomotion.

Gibberellins: An important group of phytohormones. Discovered in Japan in paddy plants infected with the fungus, *Gibberella fujikuroi.* Chemically the hormone is gibberellic acid. Today more than fifty gibberellins are known. Occur in algae, fungi, mosses, ferns, gymnosperms and angiosperms. Common in higher plants but restricted to only a few species of bacteria and fungi. Occur in the stein apex, young leaves and seeds, etc. Are capable of converting even a genetically dwarf plant into a tall plant. Gibberellins favour flowering in long day plants, break dormancy, promote seed germination, substitute cold treatment, induce parthenocarpy, etc.

Gill: (Bot.) Fingerlike spore containing structure in Agarics. (Zool.). Respiratory organ of aquatic animals, well supplied with blood; through gill surface occurs interchange by diffusion of oxygen and carbondioxide between water and blood. Gills of most fishes are internal gills within gill slits. Gills of amphibian and diphoan larvae and of adults of some urodeles are external gills, from epidermis of gillslits, i.e. ectodermal.

Gill bar: Tissue separating, adjacent gill-slits in chordates,, containing blood vessels, nerves and skeletal support.

Gill fungi: Fungi belonging to the family Agaricaceae of Basidiomycetes; possessing a characteristic fruit body consisting of a stalk (stipe) supporting a cap (pileus) on the under-surface of which are radially arranged gills (lamellae) bearing the hymenium, e.g. mushrooms, toadstools.

Gill pouch: Outpushing of side-wall of pharynx towards epidermis in all chordate embryos, precursor of gill slit in fish and some amphibia but breaks through to exterior only temporarily or never in terrestrial vertebrates. A series of gill-pouches occurs on each side, one behind the other.

Gill slit: Opening from the side of pharynx to exterior in aquatic chordates. Usually vertically elongated, several tying in a series one behind the other on each side. Primitively probably concerned in filtering food particles from water pumped through them by action of cilia (as in tunicates and amphioxus). In fish and some amphibia, serve for respiration. Absent in adult tetrapod verterbrates, except in a few amphibia, but occur in many tetrapod embryos. Presence of gill-slits or at least of gill pouches and corresponding epidermal grooves at some stage of development is a characteristic of phylum chordata.

Ginkgoales: Order of gymnospermae that flourished in the Mesozoic and now represented by one living member often referred to as a 'living fossil', *Ginkgo.*

Girdle: Skeletal structures of vertebrates (pectoral and pelvic) by which the appendages are associated with the trunk.

Gizzard: Heavily muscularized part of the alimentary canal of many animals where food is broken up. Found in crustaceans, earthworm and in birds.

Gland: An organ (sometimes a single cell) whose main function is to build up one or more specific chemical compounds (secretions) 'which are passed to the outside of the gland

Glenoid cavity: Cuplike hollow on each side of the pectoral girdle (on scapula and when present, coracoid) into which head of humerus fits, forming shoulder-joint, in tetrapod vertebrates.

Globulin: Simple proteins, which are in water, but dissolve in certain salt solutions. Present in blood may function like antibodies.

Glomerular filtration rate: The volume of plasma filtered through the glomeruli of the kidneys.

Glomerulus: Of vertebrate kidney. Small bunch of capillaries covered by thin epithelium, which projects into Bowman's capsule. Capillaries are supplied with blood by an arteriole and in birds and mammals discharge into another (efferent arteriole) which in turn supplies capillaries of uriniferous tubule. Much of the dissolved substances, and much of the water, of blood flowing through glomerulus filters into capsules and passes down tubule (where most is reabsorbed into blood stream again) Plasma proteins and blood corpuscles do not filter into capsule.

Glossopharyngeal nerve: Ninth cranial nerve of vertebrates. Controls swallowing reflex.

Glottis: The opening from the pharynx into the larynx.

Glucagon: Hormone antagonistic to insulin in action, converting glycogen into glucose in the liver to raise blood-sugar level. Secreted by alpha cells of the islets of Langerhans in the pancreas.

Glueogenic Glucose-producing, especially amino acids which after deamination metabolise like carbohydrates.

Glucosans: Carbohydrates, which give only glucose on hydrolysis and are made up of B-D-glucose only, e.g. starch, dextrin, cellulose.

Glucosuria (Glycosuria): The excretion of glucose in the urine due to high blood glucose level.

Glume: One of a pair of dry bracts at the base of and enclosing the spikelet of grasses.

Gluten: Protein contained in wheat flour.

Glycogen A soluble carbohydrate stored in muscles and liver; 'animal starch'.

Glycogenesis: Manufacture of glycogen from glucose.

Glycogenolysis: The break-down of glycogen into glucose.

Glycolipids: Complex lipids which consist of compounds of fatty acids with carbohydrates.

Glycolysis: Break down of glucose to form two molecules of lactic acid. It does not require the presence of oxygen and there is net synthesis of two ATP molecules. Enzymes for this process are located in the cytoplasm.

Glyconeogenesis: The formation of glycogen from amino acids and fats.

Glycoprotein (Mucoprotein): A protein with covalently attached sugars and polysaccharide. Present in the plasma membrane with the carbohydrate facing the extracellular surface.

Gnathostomata: The group of Chordates under the subphylum Vertebrata with true jaws, paired fins or limbs; includes fishes, amphibians, reptiles, birds and mammals.

Goblet cell: Pear-shaped cell present in columnar epithelium, the part near the free surface of the epithelium being swollen with mucin, which is secreted from time to time on the surface.

Goitre: An enlargement of thyroid gland caused by deficiency of iodine.

Golgi apparatus: A membranous organelle found in all cells except bacteria. It consists of a cluster of flattened, parallel, smooth-surfaced sacs called cisternae and many smaller vesicles. In many plant cells it consists of many unconnected units called dictyosomes. They are in dozens to hundreds. They form secretory vesicles, synthesize pectin and carbohydrate, contribute to the cell wall formation in plants.

Gonad: Reproductive organ ovary, testis) in which gametes (ova or sperms) are produced.

Gonadotropic hormones (Gonadotropins): Hormones secreted by anterior lobe of pituitary of vertebrates and by mammalian placenta, which control activity of gonads. Their output from the pituitary is controlled by the hypothalamus; changes in output are responsible for: onset of sexual maturity, for breeding season rhythms and for oestrous cycles.

Gonidium: (1) An algal cell in a lichen thallus. (2) A nonmotile spore formed by some Myxophyceae. (3) A gemma of some liverworts.

Graafian follicle: Fluid filled spherical vesicles in mammalian ova conning an oocyte attached to its wall. Differs from ovarian follicles of other vertebrates in presence of cavity. Cavity enlarges during early part of estrous cycle of those mammals which have such a cycle, separating follicle cells into an external layer and a layer adherent to oocyte. Enlargement continues until follicle bursts on to surface of ovary discharging the oocyte (ovulation).

Graft A small piece of meristematic tissue, e.g. a bud or growing shoot, led the scion, is made to unite with a larger established plant, called stock.

Gram's stain: Stain used for bacteria. Bacteria that take the stain (Staphylococcus) are Gram-positive and that do not take the stain (typhoid bacteria) are Gram-negative bacteria.

Grana: In chloroplasts, the groups of disc-shaped vesicles (thylakoids) stacked like coins. Vesicle membranes bear chlorophylls and carotenes associated with photosynthesis.

Grand period of growth: Total period of enlargement of a cell, tissue, organ or organism. Ṣtarts slowly immediately after differentiation, increases to a maximum rapidly, and finally falls off to zero.

Grassland: Region where natural vegetation is grass; used as pasture. Savannahs are the tropical grasslands.

Gray crescent: Region of presumptive chordamesoderm on the surface of egg, grey in colour, future blastopore and anus.

Grazing food chain: The food chain in which animals eat primary producers (green, plants).

Green sensitive cones: Cones which contain photopigment sensitive to light of 535 nm wavelength.

Green glands: Pair of excretory organs on each side of head region of some arthropods.

Green revolution: Spectacular increase in the yield of crop plants, especially cereals, by use of modern techniques of agriculture.

Grey matter: Tissue of vertebrate central nervous system containing numerous cell-bodies and dendrites of nerve cells; and unmyelinated terminations of nerve-fibres forming synapses with these; together with ganglia and blood-vessels. Occurs mainly as an inner layer surrounding central canal; but occurs also as superficial layer (cortex) on cerebellum and cerebral hemispheres in some vertebrates. The coordinating work of the CNS is done in grey matter by the large numbers of synapses present.

Ground tissue: Parenchyma cells in which vascular bundles are embedded in the monocot stem.

Growth: Increase in size and dry-weight of an organism or cell; which cannot be reversed. It does not include the increase in size due to the uptake of water.

Growth hormone: Hormone secreted by anterior pituitary. It stimulates growth by its action on carbohydrate and protein metabolism.

Growth promoting substance: A substance which promotes or accelerates growth. It may be formed inside the organism or may be obtained from an external source.

Growth ring: A cylinder of secondary wood laid down in a single season in a stem or root. It is seen as a circle in transverse section.

Guanine: One of the pyrimidine bases found in DNA and RNA.

Guard cells Specialised, crescent-shaped epidermal cells occurring in pairs surrounding stoma. Changes in shape of guard cells, due to changes in their turgidity, control opening and closing of stomata and hence effect loss of water vapour and gaseous exchange.

Gum arabic: Gum acacia. Gum obtained from certain varieties of acacia.

Guttation: Exudation of water from hydathodes, which are commonly at the end of the main veins of leaves.

Gymnosperms: A class of vascular plants, The seeds are naked and exposed on the sporophyll i.e., the sporophyll is not infolded to form an ovary.

Gynandromorph: An individual of a dioecious species having one part of the body female and another part male in constitution.

Gynobasic: A style which arises near the base of ovary as in Labiatae.

Gynodioecious: The condition in which female and hermaphrodite flowers are borne on separate plants.

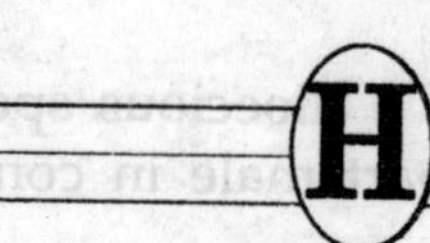

Habitat: An area with uniform climatic, vegetative, topographic, or other relevant conditions. Examples include an area of tropical rain forest or desert. Environmental conditions mavy vary within a habitat, for example in a forest the conditions at groung level are very different from those in the leaf canopy.

Hadrome: The conducting and living cells of the xylem of seed plants.

Haematology: Study of blood, its constituents and the diseases connected with it.

Haeme: One of a group of iron-porphyrins, which are conjugated with proteins to form peroxidase, catalase and all the cytochromes.

Haemocoel: Reduced coelom that also functions as part of the blood vascular system.

Haemoglobin: A red coloured iron containing respiratory pigment in the blood of vertebrates and some invertebrates.

Haemoglobinometer: Instrument for determining the percentage of haemoglobin in the blood.

Haemolysis: The destruction of the red blood cells.

Haemophilia: Human disease in which blood-clotting is defective. Known only in males Transmitted from mother to son. Determined by sex-linked recessive gene.

Hair: A slender filamentous growth on the skin of mammals.

Hallucinogen: Substance that induces hallucinations.

Halobiontic: Strictly confined to salt water.

Halophyte: Plant that can tolerate very salty conditions.

Halteres (balancers): Knob-like balancers in Diptera that represent the vestigial hind wings.

Haplodiplont: A sporophyte in which cells contain haploid number of chromosomes.

Haploid: Having only a single set of chromosomes(n) as in gametes.

Haplont Haploid stage of an organism; ending with fertilization.

Haptotropism (Thigmotropism): Tropism in which stimulus is a localised contact, e.g. a tendril in co act on one side with a solid object, a twig for instance., curves in that direction and coils round it.

Hardy-Weinberg equilibrium: The situation in a large randomly mating population in which the proportion of dominant to recessive genes remains constant from one generation to the next. It is described by the equation $p^2 + 2pq + q^2 = 1$, where p^2 and q^2 are the frequencies of the double dominant and double recessive respectively, and 2pq is the frequency of the heterozygote. The law was formulated in 1908 and disproved the then current theory that dominant genes always tend to increase in a population at the expense off their equivalent recessive alleles. The equilibrium only holds providing that the population is sufficiently large to avoid chance fluctuations of allele frequencies in the gene pool (genetic drift) and providing there is no mutation, selection, or migration. The fact that allele frequencies may be seen to change fairly rapidly in large populations that show minimal mutation and migration, emphasizes the important role natural selection must play. Until the Hardy-Weinberg law was formulated, the extent of natural selection was not fully appreciated.

Haustorium: A specialised organ of a parasite, which penetrates the host tissue, and absorbs nutrients and water.

Haversian canals: Channels (roughly 50.pm diameter) carrying blood vessels and nerves, which ramify throughout bone, communicating with its surface and marrow. Sheets of bone, and bone cells, are arranged concentrically around the canals.

Heart: A muscular organ that pumps blood around the body. In mammals, it consists of four chambers with the right and left sides totally separate. Deoxygenated blood is carried to the right atrium via the venae cavae and oxygenated blood from the lungs is carried to the left atrium by the pulmonary vein. Contraction of both atria forces blood into the respective ventricles, which in turn contract forcing blood into arteries: the pulmonary artery carries deoxygenated blood from the right ventricle to the lungs and the aorta transports oxygenated blood from the left ventricle to the body. Valves prevent the backflow of blood. The rhythmic contractions of the heart (cardiac) muscle is basically automatic (see pacemaker). In a resting man contractions occur about 72 times per minute.

The form of the heart varies greatly throughout the animal kingdom (reaching its greatest complexity in birds and mammals). In annelids there are a number of lateral contractile vessels known as hearts. Crustaceans have a single heart with openings (ostia) possessing valves. In insects the heart is a long dorsal tube divided into 13 chambers, also with ostia. Molluscs and fish have two-chambered hearts. Amphibians have two atria and one ventricle. In reptiles there is one ventricle, with a sputum that completely separates the ventricle into two in crocodiles.

Heartwood: The central layers of xylem tissue in tree trunk with no living cells; cease to conduct water and minerals. Provides only mechanical support. Darker in colour than sapwood.

Heliophyte: A plant which thrives very well in a bright sunny situation.

Helotism: A form of symbiosis in which one partner benefits more than the other e.g., the fungus in a lichen thallus benefits more than the alga.

Hemal: Pertaining to the blood or bloodvascular system.

Hemicelluloses: These are carbohydrates related to cellulose, but are mixed polysaccharides, containing other sugars besides glucose. They are easily hydrolyzed by dilute acids. Form part of cell wall and may function as reserve food in seeds, etc.

Hemicyclic: A flower in which some parts are in spirals, and others in whorls.

Hemicryptophyte: A perennial plant, usually nonwoody, with its buds at soil level. Buds are protected by the soil or surface litter.

Hemiplacenta: The chorion, the yolk sac, and generally allantois which together serve as an organ of nutritional supply to the uterine young of marsupials.

Hemiptera (Rhynchota): A large order of insects including bedbugs, cicadas, aphids and many kinds of plantbugs. Mouthparts are adapted for piercing and sucking.

Hemizygous: Genetic material that has no homologous, counterpart and is thus unpaired in the diploid state. The X-chromosome in the organisms of heterogametic sex. Hemo(hemato). A prefix-meaning 'blood'.

Hemocytometer: Instrument for counting the number of blood cells.

Hemorrhage: The loss of blood from the blood vessels.

Hensen's knot (Hensen's node): An anterior end of primitive streak.

Heparin: Substance which prevents blood clotting by stopping conversion of prothrombin to thrombin and by neutralizing thrombin.

Hepatica: Pertaining to the liver.

Hepaticae: Liverworts, a class of Bryophyta. Live in wet conditions, on soil or as epiphytes, or in water. Consisting of a thin, prostrate plant body or creeping central axis up to a few inches long, provided with leaflike expansions attached to soil by rhizoids growing from under surface. Sexual organs, antheridia and archegonia, variously grouped. Male gametes motile by flagella. Fertilization is followed by development of a capsule containing spores which, being shed, germinate to from a short, thalloid protonema from which new liverwort plants arise.

Hepatic portal system: A system of veins leading from the digestive tract to capillaries (sinusoids) in the liver of a vertebrate. Thus, absorbed products of digestion (except fat) go straight to liver, and not to heart.

Herb: Plant with no persistent parts above ground, as distinct from shrubs and trees.

Herbaceous: Having the characters of a herb.

Herbarium Collection of preserved plant specimens for identification and reference purposes.

Hereditary: Passing by inheritance from one generation to another.

Heredity: The transmission of characters, physical and others, from parent to offspring; the tendency of offspring to resemble their parents, yet allowing new types to emerge.

Hesperidium: A fleshy-fruit formed from a superior, syncarpous gynoecium. The flesh is formed from fluid-filled hairs projecting into loculi, e.g. an orange.

Heterochlamydeous: Flowers having two kinds of perianth segments (sepals and petals) in distinct whorls,

Heterochromatin: Genetically inactive part of chromatin. It is rich in RNA and is responsible for nucleic acid metabolism,

Heteroecious: Rust fungi that require two host species to complete their life cycle.

Heterogamete: Gamete differing in size and form from the other conjugant gamete.

Heterogametic sex: The sex, individuals of which have within each of their nuclei a pair of dissimilar sex-chromosomes (one-X and one Y-chromosome); or an unpaired sex chromosome (a single X-chromosome). Heterogametic sex is usually male, but is female in lepidoptera, birds, reptiles, some amphibia and fish, and a few plants.

Heterogamy: Condition in which two gametes are unlike in structure, e.g. sperm and egg.

Heterokaryosis: Presence of two or more nuclei with differing genotypes within a single cell. Fusion of fungal hyphae of differing genetic complements leads to heterokaryosis. Induced fusion of different animal or plant cells is also effective in causing heterokaryosis.

Heteromerous: A lichen thallus in which the layer of algal cells lies between two layers of fungus hyphae.

Heteromorphic: Pertains to alternation of generation, particularly in algae, meaning generations vegetatively dissimilar.

Heteroptera: A suborder of bugs having the forewings in the form of a hemelytron, i.e. having proximal half of the wing horny and the distal half membranous. Includes most plantbugs, water boatman, water scorpions and the common bedbug.

Heterosis: Hybrid vigour. Increase in vigour, size and fertility of hybrid as compared with its parents, resulting from the union of genetically different gametes, and assumed to be due to increased recombination of dominant and recessive genes.

Heterosporous: Having two types of spores, microspores and megaspores that give rise, respectively, to distinct male and female gametophyte generation; a condition present in certain pteridophyta and in spermatophyta.

Heterostyly: Condition in which the length of style differs in flowers of different plants of a species e.g., pin-eyed (long style) and thrum-eyed (short style) primroses. Anthers in one kind of flower are at same level as stigmas of other kind. Device for ensuring cross-pollination by visiting insects.

Heterothallism: A condition in algae and fungi in which sexual reproduction occurs only through participation of two thalli, each of which is self-sterile.

Heterotrophic: Organisms requiring a supply of organic material (food) from its environment. All animals and fungi, most bacteria and a few flowering plants are heterotrophic requiring organic food substance from other sources.

Heterozygote: A zygote obtained from the union of male and female gametes from two different parents and having different alleles of characters in homologous chromosomes.

Hexapoda: An alternative name for insects; so called on account of six legs.

Hexasomic: An otherwise normal diploid cell but having one chromosome represented six times.

Hexose: A sugar containing six carbon atoms.. Glucose, fructose, galactose, etc. are hexose sugars.

Hibernation: Dormancy during winter. Occurs in many mammals, most reptiles and amphibia, and many vertebrates of temperate and arctic regions. Metabolism is greatly slowed, and in mammals temperature drops to that of the surroundings.

Hill reaction: Isolated chloroplasts in suspension, when illuminated, reduce ferric ions, and produce oxygen. They cannot utilize carbon dioxide, so that it was assumed that the illuminated chloroplast their own, reduced water, the hydrogen being removed by an oxidizing agent.

Hilum: Scar on seed coat, at the point of attachment of seed to funicle.

Hindbrain: Hindmost of the three divisions of vertebrate brain. Comprises cerebellum and medulla oblongata.

Hind-wings: Of an insect. When present, are on the third thoracic segment. In Diptera, the forewings are used for flying and the hindwings are reduced to form small balancers or halteres.

Histochemistry: Study of distribution of particular chemicals by specific staining methods, etc., within sections or whole mounts of tissues.

Histogen theory: A theory in which the apical meristem is considered to consist of three main zones the dermatogen, periblem and plerome, that differentiate into epidermis, cortex and stele, respectively. This concept has now been replaced, for stem apices, by the tunica corpus theory. In roots, however, the concept is still applied, and in some angiosperm roots, a fourth histogen zone is recognized, the calyptrogen, which gives rise to the root cap.

Histology: Study of the structure of tissues and organs of living beings.

Histones: A class of basic proteins with an unusually large proportion of the basic amino acid sarginine and lysine in their makeup, associated with DNA in the chromosomes of eukaryotic cells. Probably concerned witn shutting off gene activity. Cause supercoiling of chromatin threads.

Holarctic: Zoogeographical region amalgamating the Palaearctic and Nearetic.

Holoblastic: Whole zygote undergoing cleavage. Occurs in eggs with little or moderate amount of yolk, e.g. frog.

Holocarpic: Fungi having mature thallus converted in its entirety into a reproductive structure.

Holocene (recent): Geological period consisting of recent times, since the end of the last ice-age.

Holocoenotic principle: Principle that environment acts as a whole because of lack of barriers to the interaction of its component factors.

Holoenzyme: A complete enzyme, made up of an apoenzyme and a prosthetic group.

Hologamy: The fusion of two mature cells, each of which has been completely changed into a gametangium.

Holophytic nutrition: Nutrition involving photosynthesis from simple inorganic chemical substances as in green plants and some flagellate protozoans.

Holozoic nutrition: Nutrition requiring solid organic foodstuff characteristic of most animals.

Homoestasis: The maintenance of a constant internal environment by an organism. Any deviation from this balance results in reflex activity of the nervous and hormone systems, which tend to negate the effect.

Hominid: A humanlike (opposed to apelike) creature living or extinct.

Homochlamydeous: Flowers having perianth segments of one kind (sepals) in two whorls.

Homodont: Having all the teeth of the same kind in most vertebrates, other than mammals.

Homogametic sex: The sex, individuals of which have within each of their nuclei a pair of similar sex chromosomes.

Homogamy: Condition in which male and female parts of a flower mature simultaneously.

Homoiothermal: Having constant body temperature, often maintained above that of the environment; characteristic of birds and mammals. Warmblooded.

Homokaryon (Homocaryon): A cell with more than one nucleus of identical genetic constitution For example, a fungal hypha or mycelium may have more than one such haploid nuclei in its cytoplasm.

Homologous chromosomes: Chromosomes having identical loci. They are paired in a diploid nucleus. Bear corresponding genes for same set of traits.

Homologous organs: Organs which have same basic structure and origin but differ in external appearance and function.

Homomorphic: Chromosome pairs which have the same form and size.

Homonym: The same specific name for the same plant when it is placed in another genus.

Homophyllous: Having foliage-leaves all of the same kind.

Homoplastic: Of the same structure and manner of development, but not descending from a common source.

Homosporangic: Having only one kind of sporangium.

Homosporous: Having one kind of spore that gives rise to gametophyte generation bearing both male and female reproductive organs. Characteristic of certain members of pteridophyta.

Homostyly: A species in which flowers have styles of one length.

Homothallism: In fungi when sexual reproduction takes place in a colony derived from a single spore Each thallus is self-fertile.

Homozygote: A zygote obtained from the union of male and female gametes which have identical genes in homologous chromosomes. Also homozygous for a particular character.

Homozygous: Having identical genes at the same locus on each member of a pair of chromosomes which are homologous.

Hormones: Organic substances produced in minute quantity in one part of an organism and transported to other parts where these exert a profound effect. Plant hormones play a prominent role in plant growth. In animals hormones are usually secreted by endocrine glands directly from gland cells into blood stream.

Host: Organism infected by a parasite. Definitive host is that in which the animal parasite attains sexual maturity. Intermediate host is the one which is essential for the fife cycle of an animal parasite, but in which it does not become sexually mature.

Humerus: Bone of proximal part of tetrapod forelimb (bone of upper arm of man).

Humus: Complex organic matter resulting from decomposition of plant and animal tissue in the soil of great importance for plant growth. Colloidal, improving texture and water-holding capacity of soil and forming reservoir of mineral nutrients which are absorbed by humus and prevented from being leached away.

Hybrid: The offspring of two related individuals of the same species, race or variety.

Hydathode: A water pore of water secreting gland found on the edges and tips of leaf vein, of many plants.

Hydrase: An enzyme which can add or remove water, with hydrolysis.

Hydrolase: An enzyme that catalyst a hydrolysis reaction. Digestive enzymes are good examples. Hydrolases play an important part in rendering insoluble food material into a soluble form, which can then be assimilated.

Hydrolysis: Decomposition of a substance by the insertion of water molecules between certain of its bonds, as in the hydrolysis of disaccharides into monosaccharides.

Hydrophilly: Pollination in which water carries the pollen from anther to stigma. It occurs in some pondweeds.

Hydrophyte: Plant adapted to living in water, especially a vascular plant.

Hydroponics: The growth of plants in water culture solutions rather than soil. The solutions contain the correct balance of all the essential mineral requirements. The method has been used commercially" though is not widespread. Support may be provided by using beds of gravel through which the aerated solution is pumped.

Hygrophilous: Plants which need large supply of moisture to grow.

Hymenium: The outer layer of gill bearing spores in mushrooms.

Hymenoptera Order of bees, wasps, ants, sawflies etc. insects having two pairs of membranous wings, ovipositor adapted for sawing, piercing or stinging. Social insects like ants and wasps.

Hyoid arch: Visceral arch next behind jaws of vertebrates. Dorsal part form shyomandibula; ventral part in adult tetrapods forms hyoid bone, usually supporting tongue. Contains facial nerve and forms many muscles of the face.

Hyostylic jaw-suspension: Found in most fishes. Upper and lower jaws on each side are attached at their hinge to one end of the hyomandibula, the other end of which is attached to the neurocranium.

Hyoscyamine: Alkaloid obtained from henbane, used as a sedative and anti-spasmodic.

Hyperdiploidy: Condition where full diploid complement of chromosomes is present as well as a portion of one chromosome which has become translocated.

Hyperglycemia: An excessive amount of glucose in the blood.

Hypermetropia: Long-sightedness, convex lens is used to correct this disorder.

Hypermorph: A mutant gene, having a similar, but greater effect than the nonmutant allelomorph.

Hyperparasite: A parasite that lives on or in another parasite, e.g. Ichneumon fly.

Hypertension: An abnormally high tension, especially that of blood pressure.

Hypertonic solution: A solution having higher concentration of solute molecules and lower concentration of solvent (water) than that of a solution with which it is being compared (hypotonic). When separated by semipermeable membrane, water moves by osmosis into a hypertonic solution from a hypotonic solution.

Hypertrophy: Abnormal increase in size of tissue or organ by increase in size of individual elements (cells or collagen fibres) without increase of their number, e.g. in exercised muscle; uterine muscle during later part of pregnancy; in plants, in response to certain parasites.

Hypha: An individual filament of a fungus thallus forming its vegetative body.

Hypodermis: The outemost layer of cortex, next to epidermis.

Hypogeal germination: When germination takes place, cotyledons stay below the ground, as in gram.

Hypoglycemia An abnormally low percentage of glucose in the blood.

Hypogynous: Of a flower in which the other parts arise below the gynoecium Gynoecium superior.

Hypomorph: Of a mutant gene having an effect similar to but less than that of nonmutant allelomorph.

Hyponasty: Lower side of an organ growing more quickly than the upper resulting in an upward bending.

Hypophysectomy: Surgical removal of pituitary gland.

Hypoplasia; A state of having growth of an organ less than others. Developmental deficiency.

Hypoploids: Diploids lacking a piece or pieces of chromosomes from their compliment.

I

Identical twins: Monozygotic twins, formed from the same egg. Are of the same sex.

Idioblast: Any cell that varies in cell content and wall thickness from its neighbours.

Idiochromatin: A substance within the nucleus of a cell which controls the reproduction of cell.

Idioplasm: Equivalent to germplasm of Weismann.

IgG: Gamma globulin: The main immunoglobulin in man.

IgM: An immunoglobulin found in all the vertebrates. It is of much higher molecular weight than IgG.

Imbibition: This takes place when a solvent enters a colloid, between the free capillary spaces, and the intermolecular spaces. It causes the colloid to swell. The swelling causes considerable pressure.

Implantation: Adhering of the fertilized ovum to the uterine wall.

Impulse: The message (electrochemical) conducted along a nerve fibre, Inbreeding. Breeding between closely related individuals. The most extreme form of inbreeding is self-fertilization, which occurs in some plants. It increases homozygosity.

Incipient plasmolysis: The condition when about half the cells of a tissue are plasmolysed: In this condition the osmotic pressure of cell sap is usually measured.

Incomplete dominance: System in which interaction or alleles in hybrid results in a phenotype that is intermediate between the pure dominant and pure recessive individuals.

Indefinite: Not ending in a flower, and so theoretically capable of further elongation. Racemose.

Indehiscent: Said of fruits, fruit-bodies etc., which do not open on ripening to disperse their contents.

Independent assortment: Mendel's second law. When more than two pairs of genes enter a cross, each is assorted independently of the other.

Indicator Plant: A plant that grows under special conditions of climate, or in a particular soil, or in a particular community, and thus, by its presence: indicates the general nature of habitat.

Induced mutation: Mutation as a result of man-made factors.

Inductor: A loose word which includes organizer and evocator.

Indusium: Membranous outgrowth on the underside of leaf which in ferns covers and protects a group (sorus) of developing sporangia.

Infarct: Area of dead tissue resulting from ischemia due to diminished blood circulation in the area.

Infect: To enter and establish a pathogenic relationship with a host; to enter and persist in a carrier.

Inferior ovary: When receptacle completely encloses ovary and other flower parts arise from receptacle above.

Inferior vena cava: Large vein that carries deoxygenated blood from lower half of the body to right auricle.

Infest: To occupy and cause injury to either a plant or to soil, or stored products.

Inflammation: Response of a tissue to injury, characterised by increased blood flow, redness, accumulation of leucocytes and pain.

Inflorescence: The flowering shoot, as distinct from the vegetative region. According to the type of branching, it is of two major types indefinite or racemose, and definite or cymose.

Infructescence: The inflorescence after the flowers have fallen and the fruits are formed.

Infundibuliform: Tubular below, gradually opening upwards, i.e. funnel shaped.

Ingest: To take food into the digestive tract.

Inguinal canal: Tubular opening in lower abdominal wall containing spermatic cord in males and uterine round ligaments in female.

Inheritance: The sum of all characters that are transmitted by the germ cells from generation to generation.

Inhibitor: A substance which limits or destroys the catalytic activity of an enzyme.

Initials: In meristem, cells which perpetuate themselves by division and at the same time form new body cells.

Innate: (1) Sunken into the thallus. (2) Originated in the thallus. (3) Said of an anther which is joined to the filament only by its base.

Insecticide: A toxin effective against insects.

Instar: Period between consecutive molts in insect development

Instinct: An inherited type of action, invoked by a certain stimulus and often of complex nature, combining associated reflex acts and leading to a particular response.

Insulin: Hormone secreted by beta cells of the islets of Langerhans (Pancreas). It promotes the conversion of blood glucose into tissue glycogen used in the treatment of diabetes mellitus.

Integument: The membrane (s), enclosing the nucellus, finally forming the testa (seed coat).

Intercalary meristem: Meristematic tissue situated between mass of permanent (nondividing) tissue.

Interchromocentre: The areas of a chromatid composed of euchromatin.

Inter-fascicular cambium: Cambium between the vascular bundles in stems. It is formed when parenchyma cells resume meristematic activity.

Interferon: A protein produced in animal cells as a defence against virus. It inhibits replication of virus.

Interneuron: Neuron whose cell body and axon lie entirely within the central nervous system.

Internode: The stem between two successive nodes.

Interphase: The resting stage that occurs between the first and second cell division. Physiologically, this is the most active phase of cell.

Interstitial cells: Testosterone-secreting cells, which lie between the seminiferous tubules in the testes.

Interstitial cell stimulating hormone (ICSH): A gonadotropic hormone of the anterior pituitary which stimulates androgen secretion by the Leydig cells.

Intervarietal: Said of a cross between two individuals of the same variety of the same species.

Interxylary phloem: A strand of secondary phloem surrounded by secondary xylem.

Intine: The inner layer of the wall of the pollen grain. Intracellular. Inside the cavity of the cell.

Intracellular digestion: Digestion inside the body of the cell.

Inulin: A carbohydrate compound clinically used to measure glomerular filtration rate.

Inversion: The breaking and reuniting of the parts of a chromosome so that the genes are lying in the reverse order.

Invagination: Folding of a layer of tissue to form a sac, as during gastrulation.

Invertebrate: An animal without a dorsal column of vertebrae.

In vitro: Meaning 'in glass'. Applied to biological processes made to occur in isolation from the whole organism in glass tubes or flasks.

In-vivo: Within the whole living organism.

Involucre: A protective envelope which encloses reproductive organs. Involution. The rolling inward or the turning in of cells over a rim.

Iris: The pigmented diaphragm directly in front of the lens of the eye. Its muscles by contraction adjust size of the pupil.

Irregular: Not divisible, into halves by an indefinite number of longitudinal planes. Having members of a whorl which are not all alike.

Ischemia Reduced blood supply to an organ.

Isoenzyme: Variants of same enzyme within an organism, with same properties but with slight differences in molecular structure.

Isoelectric point: p^H value at which a substance, e.g. a protein solution is electrically neutral.

Isogamete: One of a pair of uniting gametes of similar size and form, but may be physiologically different.

Isogamy: Fusion of two similar gametes.

Isogenic: Propagating entirely by means of apogamy.

Isokontan: Bearing two (or more) flagella of equal length.

Isolecithal: An egg in which yolk is not abundant and nearly uniformly distributed throughout.

Isomorphic: A condition seen in certain algae in which the alternating generations of the life cycle are morphologically identical.

J

Jaundice: Condition in which skin becomes yellowish due accumulation of bilirubin in blood as a result of its failure to be excreted by the liver into bile.

Jejunum: Part of small intestine between duodenum and ileum in mammals. It has larger villi and is the main absorptive region.

Joint: Place of union of two bones or other hard plates.

Jugular: Main vein returning blood from head region in vertebrates.

Jungermanniales: One of the orders of liverworts (Hepaticopsida). Thallus is differentiated into "stem" and "leaf" like structures. Sex organs are typically antheridia and archegonia. The capsule has more than one layered thick jacket. Includes genera *Pellia, Porella, etc.*

Jurassic: The middle period of the Mesozoic, 1901-35 million years ago. During Jurassic dinosaurs were becoming large and abundant and bony fishes (teleosts) were also evolving rapidly. Fossils of the earliest known bird *Archaeopteryx* and of the first mammals belong to late Jurassic.

Juvenile form: A young plant which has leaves and other features different from those of a mature plant of the same species.

K

Karyokinesis: Division of nucleus as distinguished from the division of the cell or cytokinesis.

Karyology: Study of nuclear structure and function.

Karyolymph: The matrix lying in the reticulum of nucleoplasm.

Karyomere: A swollen condition sometimes seen in chromosomes towards the end of nuclear division.

Karyosome: Nucleus composed of chromatin.

Karyoplasmatic ratio: Ratio of the volume of the nucleus and that of the cytoplasm of the same cell.

Karyotype: Character of a nucleus as defined by the size, shape, and number of the mitotic chromosomes.

Keel: (Carina). Thin platelike projection of bone from ventral surface of breastbone (sternum) of birds and bats, to either side of which powerful wing muscles are attached.

Keratin: Tough fibrous protein occurring in epidermis of vertebrates forming resistant outermost layer of skin, and also hair, feathers, horny scales, nails, claws, hooves, and outer coating of horns of cows, sheep, etc.

Ketosis: Presence of ketones in the body in excessive amounts.

Labium: (Bot) Lower lip of flowers of family Labiatae. (Zool) Lower lip of insects. Consists of paired appendages of one segment fused together in mid-line.

Lac: Resin exudation covering females of insect *Tachardia.*

Lachrymal glands: Tear secreting glands.

Lactase: Enzyme which catalyses the conversion of lactose into glucose.

Lactation: The production and secretion of milk.

Lacteal: Lymph vessel in a villus of mammalian intestines.

Lactiferous hyphae: (Lactiferous tubes). These are tubes occurring in some fungi. They are filled with latex which coagulates on around.

Lacuna: A space within bone or cartilage tissue where osteocytes or chondrocytes are located.

Lagging: Slow movement towards the poles of the spindle by one or more chromosomes in a dividing nucleus, with the result that these chromosomes do not become incorporated into a daughter nucleus.

Lamarckism: The theory: that acquired characters are inherited during sexual reproduction. Proposed by Lamarck.

Laminarin: A polysaccharide storage-product found in brown-algae.

Lampbrush chromosome: Long chromosome and covered by very fine hairlike projections which are actually loops—each loop with a core of DNA.

Larva: An immature stage of an animal, dissimilar from adult in appearance. Capable of independent living. After metamorphosis changes into adult.

Latent period: Interval between the application of a stimulus and first detectable response in an irritable tissue.

Lateral geniculate body: One of a pair of brain centres where the axons of the optic nerve synapse with interneurons leading to the visual cortex.

Laterite: Reddish infertile tropical soil. Because of intense leaching out of silica, only iron and aluminium rich clay is left. Becomes fertile only when a considerable amount of organic matter is added continuously.

Latex: Liquid found in some flowering plants, contained in special cells or vessels called laticifers. Latex consists of many substances in solution and colloidal suspension, some of which are important in medicine and industry.

Laticifers: Latex containing structures found in certain plants, e.g., rubber, poppy, and euphorbia.

Latimeria: The only known huge sized, living representative of the great group of fossil-fish, Crossopterygii. A coelacanth. Discovered first by Miss Latimer off the shores of Africa.

Leaf: A flattened appendage of the stem that arises as a superficial outgrowth from the apical meristem.. Leaves are arranged in a definite pattern, have buds in their axils, and show limited growth.

Leaf sheath: The lower part of leaf which covers the stem more or less completely.

Lectotype: Specimen selected from original material to serve as the type specimen when this was not designated at the time of publication or is missing.

Legumes: Plants of pea family.

Leibig's law of minimum: Amount of plant growth regulated by the factor present in minimum amount, and rises or falls accordingly, as this is increased or decreased in amount.

Lens: Of eye of vertebrates, transparent structure just behind pupil, lying in aqueous humour, attached by collagen fibres to ciliary body. In land-living forms, including man, not as important as cornea in the refraction which produces image on retina; function is accommodation by change of shape. Consists largely of elongated inert cells (fibres which contain large amounts of special lens proteins, the crystallins.

Lenticel: Small lens-shaped pore in the periderm of stems and roots packed with loose corky cells. Allows exchange of gases between the plant and atmosphere.

Leptosporantiate: Sporangia in vascular plants arising from a single parent cell and possessing a wall of one layer of cells. Spore production is low in comparison with eusporangiate type.

Lethal gene: Gene which kills the individual bearing it.

Leucoplast: Colourless plastid.

Leukemia: Blood cancer; cancer characterized by an uncontrolled increase in the number (mitosis): leucocytes in the blood.

Leukopenia: Lack of white corpuscles in the blood.

Liana (Liane): A woody, climbing plant, found in tropical forests. Have anomalous secondary thickening.

Lichens: A group of composite plants comprising an alga and a fungus in intimate association (symbiotic).

Life cycle: Progressive series of changes undergone by an organism from fertilization to the death of that state producing the gametes which begin an identical series of changes.

Ligament: A band of collagen connecting the two bones at a joint; helps restrict movement preventing dislocation.

Ligase: Enzyme that forms recombinant DNA by linking the fragmented ends of two DNA molecules previously split by restriction enzyme.

Light phase: The phase in photosynthesis when ATP and $NADPH_2$ are formed. The process is called photophosphorylation. The energy absorbed by the chloroplasts is used to split the water molecule into hydrogen and oxygen.

Lignification: Impregnation with lignin.

Lignin: Complex organic substance deposited in the cell walls of vessels, tracheids and sclerenchyma for mechanical support.

Limiting factor: An environmental factor whose deficiency or excess is restrictive to living organisms.

Lingual: Pertaining to tongue.

Linkage: Association of two or more non-allelomorphic genes so that they tend to be passed from generation to generation as an inseparable unit and fail to show independent assortment. Separation of such genes into different chromosomes occurs from time to time as a result of crossing-over. The nearer such genes are to each other on a chromosome the less often they are separated by crossing-over and the more closely linked they are said to be. All the genes in one chromosome from one linkage-group.

Lipase: A fat digesting enzyme.

Lipid: A fatty compound of carbon, hydrogen and oxygen. Proportion of oxygen is lower than in carbohydrates. Compounds of glycerol and: fatty acids.

Liquor folliculi: The fluid of mammalian Graafian follicle into which mature ovum is freed when the tissue proligerus is broken. Contains: hormones.

Lithosphere: Solid rocky outer portion of earth.

Littoral zone: The shallow water region with light penetration to the bottom.

Locule: A chamber in an anther or in an ovary.

Long-day plants: Plants which flower in response inductively to a 'short-night'. For example, radish.

Loop of henle: Hairpin like segment of the tubule of nephron situated between the proximal and distal tubules.

LSD: Lysergic acid diethylamide. A psychedelic drug which produces behavior and symptoms as hallucinations, delusions, etc. Induces a state resembling schizophrenia.

Luciferase: An oxidizing enzyme which, acting on protein luciferin causes luminosity in glowworm and firefly, and many other animals.

Lumber: Pertaining to abdominal region.

Lumen: Cavity of a tubular organ. The space enclosed by a cell wall especially after the contents have disappeared.

Lutein: Yellow-coloured substance contained in the cells which fill the ruptured: mammalian Graafian follicle.

Luteinizing hormone (LH): A gonadotropic, hormone of anterior pituitary responsible for final maturation of follicular cells, ovulation and formation of corpus luteum.

Lyases: Enzymes which breakdown complex molecules to simpler ones, without hydrolysis, e.g. carbonic anhydrase breaks down carbonic acid to water and carbon dioxide.

Lymph: The intercellular fluid collected into the lymphatic vessels for return to the blood stream. See lymphatic system also.

Lymphatic system: System of fluid containing tubes present in the vertebrates. A blindly-ending meshwork of small tubes (lymph capillaries) permeates most tissues (not nervous system) and connects up into even larger vessels which finally join venous system usually neat the heart. Lymph is drained from the tissues into the blood by this system. Because of high permeability of lymph capillaries, large molecules from the intercellular spaces, including invading bacteria which cannot get through blood capillary walls, are carried away with the lymph. In the course of the lymphatic vessels in some vertebrates are lymph nodes which filter out and destroy bacteria and supply lymphocytes to it. Lymph moves by contractility of vessels or by squeezing of lymphatic vessels by neighbouring skeletal muscles; the larger vessels having valves which ensure flow in one direction; or (except in birds and mammals) by lymph hearts.

Macroconsumers (Phagotrophs): Large organisms, mainly animals which cat other organisms or the particulate organic matter.

Macrocyclic: Life-cycle of rusts which have two alter native hosts, and large number of spores.

Macrogamete: Female gamete, Egg cell.

Macromere: Larger blastomeres located at the vegetal pole.

Macronucleus: Larger of the two nuclei of Ciliata. It is concerned with vegetative functions.

Macronutrients: Minerals that are required in relatively large amounts for the healthy growth of plants.

Macrophagous: Organisms feeding on relatively large particles of food.

Macropterous: Large-winged: a term used with particular reference to the various types of insects such as ants and termites.

Maggot: Wormlike, small legless larva with a pointed head-end and a truncated rear, e.g. those of house flies, blue-beetles and similar Diptera.

Malacophyllous: Xerophytic plants having fleshy leaves containing much water-storage tissue.

Malt: Grain which has been allowed to germinate and then heated and dried.

Maltase: An enzyme breaking maltose into two component glucose molecules,

Mammals: Belongs to class mammalia. Warm-blooded vertebrates with hair on body and mammary glands in the female give birth to the young ones. Heart four chambered and highly developed brain.

Mammary glands Milk-secreting glands in the breast of mammals.

Mandible The lower jow of vertebrates; either jaw of an arthropod.

Mangrove: Forests of tidal regions in the tropics.

Mantle: A fold ot the dorsal part of body wall of mollusca that secretes shell.

Marginal community: Plant community bordering on community of slightly different character.

Marine: Pertaining to or inhabiting the sea or other salt waters

Marsupial: Any of the pouched mammals such as opossums, wombats and kangaroos.

Marsupium: Pouch of many marsupials and Echidna. Fold of skin supported by epipubic bones of hip girdle, containing mammary glands, into which the newborn are placed.

Mass flow: A mechanism proposed by Munch to explain the mechanism of phloem transport. The movement of substances is believed to be the result of changes in osmotic pressure.

Mast cell: Granular cell of vertebrate connective tissue that secretes histamine and heparin.

Mastigonemes: Exceedingly fine lateral projections along the length of certain flagella. Increase efficiency of flagella.

Matrix: Intercellular material in which animal cells are embedded especially those of connective tissue like bones and cartilage. Also innermost non-membranous area of mitochondria.

Maturation of germ cells: Final stages in preparation of sex cells for mating, with segregation of homologous chromosomes so that each cell or gamete contains half the usual number (due to meiosis).

Maxilla: 1. One of a pair of feeding mouthparts of various arthropods, such as crustaceans, insect, millipedes and centipedes 2. One of pair of large hones of the upper jaw of vertebrates in mammals they bear molar and premolar teeth.

Meckel's cartilage: A paired cartilage forming the lower Jaw in cartilaginous fishes such as sharks, skates and dog fish. In mammals they bear molar and premolar teeth.

Medulla: The inner portion of an organ, as opposed to the external cortex. Adrenal medulla. Medulla is the posterior region of brain. Of plants, pith central core of usually parenchymatous tissue.

Medulla oblongata: A region of the hind brain concerned with the functioning of the visceral organs e.g. stomach, lungs and heart. Leads to the spinal cord.

Medullary bundle: A vascular bundle running close to the pith.

Medullary ray: Parenchyma cells running radially through the vascular tissue of a stem or root. They are called vascular rays (secondary) when cut off by the cambium.

Medusa: Free-living stage in the life cycle of coelenterates. Reproduces sexually to give rise to polyp.

Magaphyllous: Having large leaves, especially when supplied with vascular bundle(s), in ferns and seeds plants.

Megaspore: Spore which gives rise to female gametophyte.

Meibomian glands: Sebaceous glands of the eye lids.

Meiocyte: Any cell which can undergo meiosis. Always diploid.

Meiospore: Spore produced by meiosis.

Meiosis: Cell division associated with gamete formation in which chromosome number is reduced to half. Reduction division.

Meissener's plexus: A network of nerves in the submucosa of the small intestine.

Melanin: Dark animal skin pigment in chromatophore cells known as melanocytes. Responsible for skin and hair pigmentation.

Melanocyte secreting hormones (MSH): Two types of hormones secreted by intermediate lobe of pituitary. Cause skin lower vertebrates.

Melanophore: Chromatophore in skin containing pigment melanin.

Membrane: A thin soft sheet of cells or of material 'secreted by cell.

Mendel's laws: First law of segregation, second law of independent assortment.

Meninges: Membranes covering the vertebrate, central nervous system, dura-mater outside pia-arachnoid. Single in fishes.

Meningitis: Inflammation of meninges.

Menopause: The permanent cessation of ovulation and the menstrual cycle in human female, between the age of 42-52.

Menstrual cycle: Modified oestrous cycle of certain primates (old world monkeys, anthropoid apes and man) which has special feature of sudden destruction of mucosa of uterus at the end of luteal phase of cycle, producing bleeding; and absence of any well-marked period of oestrus.

Mericarp: One-seeded portion of a fruit found in schizocarpic fruits.

Meristem: A localised group of cells, which are mitotically dividing and undifferentiated, but ultimately giving rise to permanent tissue.

Meroblastic: Incomplete cleavage of an egg in which only part of the protoplasm divides, leaving the yolk undivided; characteristic of eggs with much yolk, e.g. insects, reptiles and birds.

Merozygote: Result of sexual process in certain bacteria in which transfer of genetic material from donor to recipient of two conjugating cells is incomplete.

Mesarch: Denoting a stele or part of a stele in which protoxylem is surrounded by metaxylem.

Mesenchyme: The mesoderm cells especially abundant in embryos and in primitive adult animals, often jelly-secreting.

Mesentery: Double layer of peritoneum attaching visceral organs to dorsal wall of peritoneal cavity. Contains blood, lymph and nerve supply to the organs.

Mesocarp: Middle layer of the pericarp (fruit wall).

Mesoderm (Mesoblast): The germ layer from which muscles, connective tissue and blood system usually develop. At gastrulation the mesoderm comes to lie between ectoderm on the outside and endoderm lining the gut.

Mesoglea: Noncellular gelatinous layer between ectoderm and endoderm of coelenterates.

Mesonephros: The functional kidney of adult fishes, and amphibians.

Mesophyll: The chlorenchymatous cells of a leaf which may be differentiated into the palisade cells with a large number of chloroplasts and the spongy parenchyma of loosely packed cells with fewer chloroplasts. Lie between upper and lower epidermis.

Mesophyte: A plant occurring in places where water supply is neither scanty nor abundant.

Mesorchium: Mesentery that surrounds and supports the testis to the body wall.

Mesosome: Invagination of the plasma membrane of certain bacteria associated with respiratory enzymes and comparable (functionally) to the mitochondria of eukaryotes.

Mesovarium: Mesentery which suspends the ovary from the dorsal body wall.

Mesozoic: The middle era in the geological time scale, dating from about 225 till 65 million years ago. Known as the 'Age of Reptiles', divided into three main periods: Triassic, Jurassic and Cretaceous.

Messenger RNA (mRNA): Form of RNA that transfers genetic information from DNA to the ribosomes where proteins are synthesized.

Metabolic pathway: Sequence of enzyme mediated chemical reactions leading to the formation of a particular product.

Metabolite: Any substance which takes part in the process of metabolism.

Metabolism: Sum total of the vital activities of constructive (anabolism) and destructive (catabolism) nature in an organism.

Metacarpal bones: Bones of the forefoot of tetrapod vertebrate. Rod like bones, usually one corresponding to each digit. Articulate with carpals proximally and phalanges distally.

Metamere: Any one of the series of homologous parts in the body as with annelids, arthropods or chordates; a somite.

Metamerism: Segmental repetition of homologous parts (metameres).

Metamorphosis: Series of changes from larva to adult after embryonic development.

Metanephros: Adult kidney of reptiles, birds and mammals

Metaphase: Second stage in meioses or mitosis when the chromosomes become arranged on the equator of the spindle.

Metaphloem: Primary Phloem consisting of sievetubes, fibres and parenchyma.

Metastasis: Breaking away of cancer cells from parent tumour and their spread to other parts of body.

Metatarsal bones: Bones of foot of tetrapod vertebrates. Rod-like corresponding to each digit. Articulate with tarsals proximally and with phalanges distally.

Metaxenia: Effect that may be exerted by pollen on the tissues of the female organs.

Metaxylem: Elements of xylem that differentiate after protoxylem. Cells wider with more lignified walls.

Metazoa: Multicellular animals in which cells are organised in tissues.

Microbiology: Branch of biology dealing with the structure and function of microorganisms.

Microbodies: Cytoplasmic organelles, spherical or oval, 0. 1 to 1.5 micrometers across, bounded by membrane. Contain enzymes. Found in a number of different plant and animal cells.

Microconsumers (Saprotroph , Osmotrophs, Microorganism): Chiefly bacteria and fungi which breakdown complex molecules of dead cells and absorb these substances.

Microfibril: Fibril consisting mainly of actin and commonly present in eucaryotic cells.

Microfilament: Filament found in eukaryotic cells having globular protein subunits identical to the actin of muscle.

Microflora: The microscopic plants, e.g. fungi and bacteria of an area.

Microgamete: Male gamete: sperm cell.

Micrometre: The unit of microscopic measurement. 1/1,000 of a millimetre, represented by um

Micron: One-thousandth of a millimetre.

Micronucleus: In ciliates the micronucleus is the smaller of the two nuclei present, it divides amitotically and provides the gametes during conjugation.

Micronutrient: Minerals which are needed in relatively small amounts for healthy growth (usually spoken for plants),

Microorganism: Organism of microscopic size such as bacteria, protozoans and many algae.

Micropyle: Minute pore in seed coat through which water enters the seed during germination.

Microsome: A minute vacuolated membrane bound structure obtained after centrifugation of' E. R R. Ribsomes are attached to it.

Microtome: Apparatus for cutting thin sections of material for microscopical examination.

Microvilli: Minute fingerlike projections from a cell surface, around a tenth of a micrometre in diameter. On free surface of epithelium, constitute a brush border.

Microphagus: Organism feeding on very small particles of food.

Midbrain (Mesencephalon): Middle of the three divisions, marked out by constrictions, in the embryonic vertebrate brain during development becomes very thick-walled, with small central cavity.

Middle lamella: Layer of calcium pectate running between adjoining primary cell walls.

Mildew: A general term for a superficial growth of fungus. A plant disease caused by a powdery or downy mildew

Minamata disease: A disease first detected in coastal Minamata area of Japan. Caused by the entry of mercury in the food-web. Symptoms are weak muscles, impaired vision, mental retardation, paralysis and finally even death occurs .

Minimal medium: A culture medium which contains the minimum amount of materials for complete metabolism to take place.

Miocene: A geological period, a subdivision of the Tertiary period from 26 till 7 million years ago.

Miracidium: Ciliated larva of flukes. Emerges from egg, which is released from vertebrate host in excreta and parasitizes a snail where it reproduces asexually.

Mitochondrion: An organelle of all plant and animal cells, chiefly associated with aerobic respiration. It is surrounded by two membranes, the inner of which forms fingerlike processes called cristae which project into the gel-like matrix. They contain the enzymes etc., of aerobic respiration.

The reaction of Krebs cycle take place in the matrix and those of electron transport coupled to oxidative phosphorylation (i.e. the respiratory chain) on the inner membrane.

Mitosis: Cell division by which the cell nucleus and cytoplasm divide in two. It is divisible into karyokinesis and cytokinesis. It has four phases; prophase, metaphase, anaphase and telophase. Resulting daughter cells have the same number of chromosomes as in the cell.

Prophase

Chromosomes appear and shorten and thicken. Nuclear membrane disintegrates and nucleolus disappears

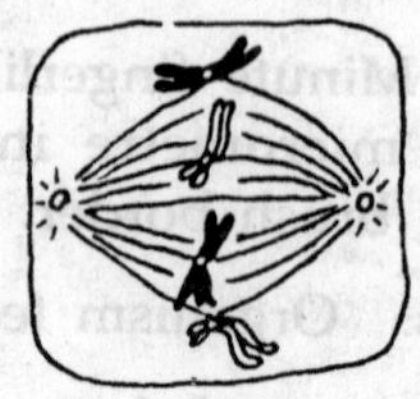

Metaphase

Individual chromosomes become aligned along the equator of the nuclear spindle

Anaphase

Chromosomes split at the centrome and the daughter chromosomes move to opposite poles of the spindle

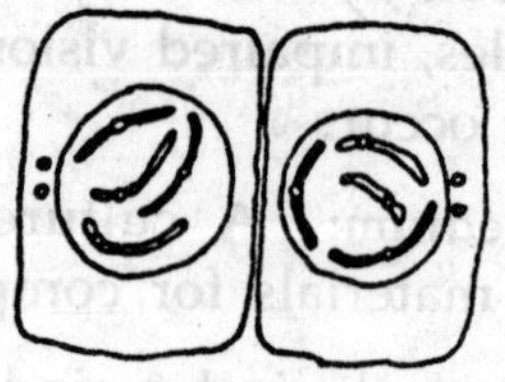

Telophase

A nuclear membrane forms around each group of daughter chromosomes

Mitral valve: Bicuspid valve between left auricle and left ventricle of mammalian heart, consisting of two membranous flaps.

Molars: Permanent crushing back teeth of mammals which (unlike pre-molars) have no predecessors in the milk teeth. They have usually several roots and a complicated pattern of ridges and projections on the biting surface.

Mollusca: Soft-bodied unsegmented animals, with an anterior head, a ventral foot and a dorsal visceral mass. Body surrounded by a fleshy mantle and protected in a calcareous shell. Includes snails, oysters etc.

Monocarpic: Plants that flower once during life-time.

Monochasium: A uniparous cyme in which each flowering branch bears another, flowering bud in its turn

Monochlamydeous: A flower with one whorl in the perianth.

Monoclinous: Having stamens and carpels.

Monoculture: The growing of a single kind of crop plants

Monocyte: Type of leukocyte that leaves blood stream and is transformed into tissue macrophage.

Monoglyceride: Glycerol carrying a single fatty acid.

Monohybrid cross: Cross in which inheritance of only one pair of contrasting character is studied at a time.

Monopodium: Axis increasing in length by the division of an apical meristem, and branching by similar lateral branches occuring in acropetal succession.

Monosaccharide: A carbohydrate composed of a single sugar molecule. For example, glucose.

Monohybrid: Offspring of parents differing in one character.

Monosomic (monosomy): Otherwise diploid organism, lacking one chromosome of its proper complement.

Monospermy: Fertilization accomplished by only one sperm. Opposite of polyspermy.

Monotremata: An order of subclass Prototheria of Mammalia including duck bill, platypus and two genera of spiny anteaters. A very primitive group of mammals laying eggs and possessing many other reptilian features. But have hair and secrete milk.

Monotypic: A species or genus which is exemplified in only one type.

Monstrosity: A marked aberrant variation suddenly appearing in species.

Morphactins: Group of synthetic plant growth regulators, derived from fluorene carboxylic acid, that reduce and modify growth.

Morphine: Alkaloid present in opium. Used in medicine for relieving pain. Dangerous habit-forming drug.

Morphogenesis: The development of structure or form of organism.

Morphology: Study of form and structure of organisms.

Morula: A spherical mass of blastomeres, without blastocoel, 16-celled in man, but usually 32 or 64-celled stage in other animals.

Mosaic vision: Mode of vision of compound eye (of insects), forming as many apposition images as there are visual units.

Motor: Pertaining to muscles, movement and stimulation of effector organs.

Motor end-plate: The junction between axon of a motor neuron and the surface of a muscle fibre. Has receptor site for acetylcholine.

Motor neuron: A neuron that conducts the impulse to an effector (muscle or gland) to elicit a response.

Motor unit: All the skeletal muscle fibres stimulated by a single motor neuron.

Moulting: The shedding and replacement of an outer covering such as feathers or exoskeleton of chitin as in insects.

Mucilage: Complex polysaccharide-related compound of plant origin. Has glue-like properties

Mucous membrane (Mucosa): Consists of surface epithelium containing goblet cells which secrete mucus and is underlaid by connective tissue. It lines the intestinal and respiratory tracts.

Mullerian duct: Oviduct of female gnathostome vertebrates. It develops in both sexes from embryonic mesoderm in association with the Wolffian duct, but becomes vestigial in the male. In most vertebrates, it is paired (single in birds). In mammals it is usually differentiated in Fallopian tube, uterus and vagina.

Multicellular: Said of an organism with more than one cell.

Multiple allele: Group of two or more alleles of one gene, only two of which can be present in a diploid cell. They originate by mutation.

Multiple fission: Sexual reproduction in some unicellular organisms in which nucleus repeatedly divides followed by division of cytoplasm, usually occurring within a cyst.

Muscle: Tissue consisting of cells which are highly contractile. Responsible for movement. Most of the edible parts of animal (flesh) is muscle. Stores fat and glycogen.

Mutagens: Factors capable of causing mutations.

Mutation: Abrupt and heritable change in a character. Also the change in a gene responsible for it.

Muton: Small genetic segment that can mutate.

Mutualism: Mutually beneficial association between indivi-duals of two different species, e.g. lichens.

Mycelium: A filamentous mass comprising the body of a fungus, each filament being called a hypha.

Mycophyta (Fungi): Mushrooms, moulds, rusts, etc. Group of simple eukaryotic or organisms lacking chlorophyll unicellular or possess tubular filaments, hyphae; reproducing asexually and sexually with formation of spores, often produced in enormous numbers. Fungi five either as saprophytes or as parasites.

Mycoplasma: Smallest free-living microorganisms. Prok-aryotic, lacking cell wall, variable in form and able to pass through bacteria retai-ning filters, Grouped in genus Mycoplasma and usually classified with bacteria. Sonic are saprophytic, others cause diseases of animals and man.

Mycorrhiza: The symbiotic association between a fungus and the roots of a plant. It is of two types: endotrophic in which fungus is within cortex cells and ectotrophic in which fung is external.

Mycosis: A disease of animals caused by invasion by fungus. For example, ring worm.

Myelinated nerve fibre: Nerve fibre that possesses a fatty sheath.

Myelin sheath: A fatty, non-cellullar layer that surrounds a nerve fibre. Insulating layer.

Myeloid tissue: Tissue pro-ducing myeloid elements (polymorphs and red blood cells) of vertebrate blood which are formed throughout life. Except in embryo, usually, located in bone marrow. In adult man, present mainly in ribs, sternum, skull, hip girdle and vertebrae.

Myoblast: Embryological cell that gives rise to muscle cells.

Myocardium: The cardiac muscle that forms the walls of heart.

Myofibrils: Contractile fibrils of filaments within muscle tissue subdivision of the muscle fibre found in the sarcoplasm, composed of the proteins, actin and myosin.

Myogenic: Originating within muscle tissue. Vertebrate heart is a myogenic heart since it has in-built mechanism of contraction when its nerve supply is cut off.

Myoglobin: Form of haemoglobin occurring in muscle fibres.

Myoneural junction: Junction between a motor neuron and a muscle fibre. Also called as neuromuscular junction.

Myopia: Nearsightedness, correction by concave lenses.

Myosin: Most abundant and contractile protein (globulin) found in muscles. In muscle contraction, myosin molecules combine with actin present in adjacent thin filaments to form actomyosin complexes.

Mytome: That part of somatic mesoderm which forms striped muscles,

Myriapoda: Class of Arthropods with no cephalization, poorly developed head, one or two pairs of legs per segment; includes centipedes (e.g. Scutigera) and millipedes (Julus). Tracheal respiration, excretion with Malpighian tubules, simple eyes.

Myxamoeba: Naked non-flagellate amoeboid cell, produced on the germination of a spore of the Myxothallophyta (slime fungi).

Myxodema: Pathological condition of the adult, associated with malfunction of the thyroid, in which the subcutaneous tissue becomes filled with a mucin-like material.

Myxobacteria: Rod-shaped bacteria that are distinguished by gliding movement in contact with solid surface and by delicate, flexible cell wall.

NADP: Nicotinamide adenine dinucleotide phosphate; a coenzyme similar in its action to NAD. Takes part in synthesis reactions.

Nanometre (rim): A unit of length equal to 101 metre. It is one thousandth of a micrometre. Narcotic. Drug that relieves pain and induces sleep. In large doses, it causes stupor, coma or death.

Naris (nares): Nostrils. The opening of the air passages, both internal and external, in the head of vertebrates.

Nasal: Pertaining to the nose.

Nasal cavity: Cavity of tetrapod head of olfactory organs communicating with mouth and with surface of head by internal and external nares respectively. Lined by mucous membrane.

Nastic movements: Paratonic variation movements of plant parts in which direction of movement is independent of the direction of stimulus.

Natural ecosystem: An ecosystem which ultimately produces the climax community, which develops in the absence of any major disturbance and consists of a complex, self-regulating food-web of several trophic levels.

Natural selection: The theory of organic evolution propounded by Charles Darwin.. Individuals of a population vary slightly. Certain variations are more favoured by environment and consequently survive and reproduce.

Neanderthal man: *Home neanderthalensis,* a rather recently (perhaps 40,000 years ago) extinct species of hominid from pleistocene (about 100,000 years ago).

Nearctic: Zoogeographical region comprising land, North America southwards to the middle of Mexico.

Neck canal cells: The cells present in the neck of an archegonium.

Necrosis: Death of a cell or group of cells.

Nectary: A glandular structure in flowers that secretes a sugary liquid.

Nemathelminthes: Cylindrical unsegmented round worms or thread worms. Triploblastic and bilaterally symmetrical. Mouth and anus both present.

Neo-Darwinism: Theory of evolution through natural selection of Darwin modified and expanded by Mendelian genetics. That answered many questions which Darwin's theory raised but could not adequately explain because of the lack of knowledge at the time it was formulated.

Neo-Lamarckism: Various modified versions of Lamarck's theory of evolution, sometimes incorporating the ideas of natural selection, though generally still insisting that inherited changes can be related directly to particular environment.

Neolithic: Phase of human prehistory when plants were cultivated and animals domesticated. Started about 10,000 years ago.

Neopallium: Roof of cerebral hemispheres of vertebrate brain not connected with sense of smell; serves more general coordination. Main mass of cerebral cortex of man.

Neoteny: Retention of larval or other juvenile features beyond the normal stage in the development of an animal. For example, in some salamanders like Ambystoma (Axolotl larva). Proteus, *Necturus and* Siren.

Nephridium: A tubular excretory organ in many invertebrates like Annelida, Mollusca and Amphioxus.

Nephron: Structural and functional excretory unit of vertebrate kidney, consisting of a Malpighian capsule and renal tubule. Filters blood to form urine. About 1 million per kidney in man.

Nephrostome: Ciliated entrance from the coelomic cavity into a nephridium or kidney tubule.

Nerve: Bundle of motor and or sensory nerve-fibres with accompanying connective tissue and blood vessels, in a common sheath of connective tissue. Mixed nerves, such as the spinal nerves, contain both sensory and motor fibres. Each fibre conducts its impulse independent of other fibers: in a nerve.

Nerve cord: Solid strand of nervous tissue, forming part of nervous, system of invertebrates.

Nerve fibre: The axon of a neuron.

Nervous system: A control and coordinating system of all multicellular organisms except sponges. All nervous activity is based on reflexes, the operational units of nervous system. These are routed through reflex areas comprising receptor, sensory pathway, modulator, motor pathway and effector.

Nervous tissue: Nerve cells and/ or their nerve fibres together with accessory cells which closely surround them e.g. Schwann cells of vertebrate nerve fibres, and supporting connective tissue or glia with blood-vessels.

Net primary production: Gross primary production minus the amount of chemical energy used in respiration.

Neural canal: Canal within the vertebral column through which runs the spinal cord

Neuroblast: Embryonic cell that develops into a nerve cell.

Neurohumor: Substance released at the synaptic knob of a chemical agent of transmission and excitation that either stimulates or inhibits the next neuron or muscle fibre. Acetylcholine and noradrenaline are important neurohumors.

Neurohypophysis: The posterior lobe of pituitary.

Neuromotor apparatus: Found in motile plants, gametes and zoospores, of the Chlorophyta. Intimately associated with nucleus and is responsible for controlling the movement of flagella.

Neuro-muscular junction: Junction between a motor neuron an muscle fibre. Also called myoneural junction.

Neuron: A nerve cell. The functional unit of nervous system. A neuron is composed of a cell body (cyton), an axon and one or more dendrites over which nerve impulses pass.

Neuropore: An opening into the neurocoel due to a lag in the fusion of neural folds at the anterior extremity.

Neurotransmitter: Chemical agent released by one neuron which acts upon a second neuron or a muscle or a gland cell and alters its electrical activity.

Niacin: Nicotinic acid. Its deficiency causes pellagra. Vitamin of B group.

Niche: Denotes the status of a plant or animal in the community i.e. its biotic and trophic relationship with other members of the community.

Nicotine: Alkaloid present in tobacco leaves.

Nicotinamide Adenine Dinucleotide Phosphate (NADP): Complex organic compound which helps in the transfer of hydrogen in the hydrogen transport chain before the latter combines with oxygen to form water during cellular respiration.

Nictitating membrane: Third eyelid, transparent fold of skin, lies at inner (anterior) corner of eye or below lower eyelid. It opens and closes laterally across cornea. Occurs in some sharks and amphibia, and widespread in reptiles and birds. Well-developed in a few mammals like whales.

Nidicolous birds: Birds that hatch helpless, featherless eye-closed, and in relatively underdeveloped state and stay in nest for some time after hatching.

Nidifugous birds: Those birds which hatch well-developed young, ones, and leave the nest immediately.

Nitrification: The oxidation of nitrogen containing ions in the soil, by bacteria, Nitrosomonas oxidises ammonium ions to nitrites and *Nitrobacter oxidises* nitrites to nitrates.

Nitrogen cycle: By the action of ammonifying bacteria organic matter is converted to ammonium compounds during decay in the soil. These are converted to nitrates which are utilized again by plants. Some of the nitrates are converted to gaseous nitrogen by denitrifying bacteria. But this loss is counterbalanced by the fixation of nitrogen by bacteria and certain blue-green algae.

Nitrogen fixation: The symbiotic bacteria, *Rhizobium* species, are associated with leguminous plants, forming the nodules on their roots The bacteria convert atmospheric nitrogen to inorganic nitrogen compounds while the legume supplies the bacteria with carbohydrates Free-living bacteria that can fix nitrogen includes members of the gen. era Azotobacter and Clostridium. Some sulphur bacteria (e.g. *Chlorobium),* some blue-green, algae (e.g. Anabaena), and some yeast, fungi have also been shown to fix atmospheric nitrogen. This is a very important step of nitrogen cycle.

Node: That part of stem at which one or more leaves are attached.

Nodes of Ranvier: Constric-tion of neurolemma occurring at interval along the peripheral myelinated nerve fibres.

Non-conjunction: The failure of synapsis to take place between homologous chromosomes.

Non-disjunction: The failure of homologous chromosomes to move to separate poles, both homologues going to a single pole, during anaphase I. This results in two of the four gamates formed missing chromosome.

Non-sense codons: Codons which do not code for any amino acid. They are thought to terminate messages.

Noradrenaline (Norepine-phrine). Lacks a methyl group present in adrenaline.

Notochord: Skeletal rod of large vacuolated cells lying between C.N.S. and the gut. Present a, come stage of development of all chordates. In most vertebrates occurs complete only in embryo, later the vertebrae replace it, Occurs in larval and adult amphioxus, in larval tunicates and is represented in adult hemichordata.

Nucellus: The central tissue of ovule, containing the embryc sac and surrounded by the integument. Megasporan-gium.

Nuclear genes: Genes that occur in the nucleus, in contrast to plasmagenes in cytoplasm.

Nuclear membrane: Consists of two layers; the outer one being continuous with the endoplasmic reticulum. It is perforated by round holes called nuclear pores.

Nuclear pores: Openings in nuclear membrane through which RNA Moves.

Nuclear sap: The non-basophilic fluid part of nucleus.

Nucleases: Enzymes inducing hydrolysis (breakdown) of nucleic acid,

Nueleic acids: Polynucleotides having nucleotide units which consist of alternate units of phosphate and a pentose sugar, which has a purine and pyrimidine base attached to it

Nucleo-cytotoplasmic ratio: Ratio of nucleus volume to cytoplasm volume. Alteration of this ratio may have occur because of cell-division. Increases during cleavage.

Nucleoid: Nuclear material of bacteria and blue-green algae.

Nucleolus: A more or less spherical structure found in nuclei. Rich in RNA.

Nucleoplasm: Fluid matrix of nucleus in which nucleoli and chromatin are suspended.

Nucleoprotein: Compounds of nucleic acids and proteins.

Nueleoside: A molecule consisting of abase (purine or pyrimidine) and a sugar (a ribose or deoxyribose).

Nucleosome: Beadlike repeating units in the chromatin observed under electron microscope. An oblate particle about 50-55 Å high and 110 Å in diameter.

Nucleotide: Sub-unit of nucleic acids, DNA and RNA. A chemical unit consisting of a purine or pyrimidine base, ribose or deoxyribose sugar, and one molecule of phosphoric acid.

Nucleus: The chief organelle of eucaryotic cell. It is bounded by a double-layered nuclear membrane, contains chromosomes; controls the activities of the cell; determines the transmission of inheritable characters as a result of division of chromosome.

Nucleus of Pander: Plate of white yolk below blastodisc of the chick egg.

Nudation: The formation of an area devoid of plants; by natural or artificial means. Prerequisite of plant succession.

Nulliplex: A condition of a polyploid in which all chromosomes of one homologous type carry the recessive allelomorph of a particular gene.

Nullisomy: Absence of a pair of chromosome from the somatic set.

Nut: A hard, dry, usually one-seeded indehiscent fruit, derived from a syncarpous ovary. Pericarp hard and woody.

Nutation: Autonomic growth movement in which the apex of a plant organ shows spiral growth.

Nutrient cycling: Pathway of nutrient substances from their occurrence it, the physical environment to their incorporation in living organisms and their return to the physical environment through metabolic activity death and decay of organisms,

Nyctalopia: Night-blindness; a disease of the eyes caused due to the deficiency of vitamin A. Due to damaged rods, vision during night is not possible.

Nyctinasty: The paratonic movement of variation of plant organs due to periodic alternation of day and night; e,g. opening and closing of flowers.

Nymph: Immature stage of an insect with incomplete metamorphosis.

Obesity: Excessive accumulation of fat in the adipose tissues of body. Obligate parasite. A parasite which is incapable of free existence, e.g. liver fluke, tapeworm, etc.

Obligate saprophyte: A saprophyte which cannot survive in the absence of dead organic matter.

Occipital: Pertaining to back part of head.

Occipital condyle: Bony knob at the back of skull, articulating with first vertebra.

Occiput: The Posterior portion of Vertebrate Skull where it joins the vertebral column. A skeletal plate at the back of insect head.

Oceanic: Inhabiting sea, deeper than 200 metres.

Ocellus: Simple eye of an arthropod, an eye spot.

Ocular: Pertaining to the eye.

Oculomotor nerve: Third cranial nerve of vertebrates. Almost entirely motor and supplying four of the extrinsic eye muscles, the inferior Oblique muscle, and all rectus muscles except external. A ventral root

Odontoblasts: Cells that secrete the dentine of a tooth.

Oesophagus: The tube that connects the Pharynx to the stomach.

Oil cake: Mass of oilseeds left after oil has been extracted. Used as cattle feed.

Olecranon Process: Bony process on ulna of mammals extending beyond elbow joint, for attachment of muscles which straighten

Olcosome: A large fatty inclusion in the cytoplasm of a cell.

Olfactory: Pertaining to the sense of smell.

Oligocene: Epoch of the Tertiary period of the Coenozoic era, from 3826 million years ago.

Omentum: A fold of peritoneum connecting stomach with other abdominal viscera.

Ommatidium: Visual unit in a compound eye.

Omnivorous: Feeding upon both plants and animals.

Oncosphere: Six-hooked embryo of tapeworm.

Oncogenic: Cancer-causing.

Oncology: Scientific study of cancer.

Ontogeny: Stages of developmental changes of an organism.

Oocyte: An egg mother cell in the ovary of an animal that gives rise to an ovum.

Oogamy (Anisogamy): The fertilization of a large non-motile female gamete by a small motile male gamete.

Oogenesis (Ovogenesis): Process of formation, maturation and transformation of primordial germ cells to the mature ova.

Oogonium (Bot): Female sex organ of certain algae and fungi containing one or more eggs. (Zool). Cell of animal ovary which undergoes repeated mitosis to form oocytes.

Ooplasm: The central portion of oogonium of some oomycetes,, which is more or less the undifferentiated egg.

Oosphere: The large, naked non-motile female gamete.

Oospore: Thick-walled resting zygote formed from fertilized egg in algae fungi arid other plants.

Operculum (Bot.): Characteristic of mosses. It covers the capsule of peristome arid may be forcibly blown off the capsule by pressure developing in the lower portion of the capsule. (Zool.) Covering-plate over gill-slits in bony fishes. Calcareous plate closing the opening of the shell in snails. Also lid in the egg capsule of liverfluke.

Operator gene: Initial gene of an operon which controls transcription of genes in operon. The site of repressor binding,

Opium. Dried, milky juice from unripe fruits of poppy plant (*Papaver sominiferurn.*)

Operon: A set of genes, meant for regulating synthesis of enzymes. The regulator gene produces a substance that binds with the operator, renders it inoperative and so prevents enzyme production. Presence of a suitable substrate prevents this binding and so enzyme production can commence, Another site in the operon, the promoter, initiates the formation of the messenger RNA that is synthesized by structuralgene

Opisthocoelous: Concave behind, as the centrum of some vertebrae.

Opthalmic: Pertaining to the eye.

Opsin: The protein part of the visual pigments of eye. Rods and three types of cones have different opsins.

Optic: Pertaining to the sense of sight.

Oral: Pertaining to or near the mouth.

Ordovician: The second oldest period of Palaeozoic era, some 500 till 440 million years ago.

Organ: A part of an organism that is made up of a number of different tissues specialized to carry out a particular function. Examples include lung, stomach, wing, leaf, etc.

Organ of Corti: Collection of structures in inner ear which are capable of transducting the energy of waves in' a fluid medium into electrochemical energy of the action potential in the auditory nerve.

Organ systems: Collection of organs which together perform an overall function. For example, digestive system.

Organelle: Metabolically active structure in the cytoplasm, performing specialised function.

Organism: A living plant or animal.

Organogenesis: Formation of organs during development.

Ornithine cycle: Probable method of urea formation in ureotelic vertebrates. Carbon dioxide and ammonia combine with amino acid ornithine to give the amino acid, arginine, which is split by enzyme arginase into ornithine and urea in the liver.

Ornithology: Study of birds.

Orthogenesis: A theory of organic evolution according to which variation is determined by the action of the environment on the fixed constitution of an organism.

Orthogeotropism: Paratonic growth movement (tropic) in plants in which stem grows vertically upwards and root grows vertically downwards due to gravity.

Orthotropous (Atropous): An ovule which is straight, i.e. with the micropyle in a straight line with the chalaza and funicle.

Osculum: Opening on the body of' sponge through which water is expelled.

Osmometer: Instrument that is used to measure osmotic pressure.

Osmoregulation: Control of osmotic pressure within an organism.

Osmosis: The movement of water from a dilute solution to concentrated solution through a semipermeable membrane.

Osmotic pressure: Force that a dissolved substance exerts (due to tile motion of its molecules) on a semipermeable membrane through which it cannot pass.

Ossicle: A small bone, such as those that transmit vibrations through the middle car, e.g. malleus, incus and stapes in mammals.

Ossification: Formation of bone, in both evolutionary and embryological sense.

Osteoblasts: Cellsresponsible for the formation of calcified intercellular substance of bone. Cells within bone (osteocytes) are osteoblasts which have become included as the bone developed, and have ceased to divide and form bone substance.

Osteoclasts: Multinucleate Cells which break down the calcified intercellular substance of bone. Stimulated by parathyroid hormone. Remodelling of bone shape by such breaking down constantly accompanies bone growth.

Osteogenesis: Formation of bone in embryo or during regeneration.

Ostiole: The opening in certain algae and fungi by means of which gamete or spores escape from a conceptacle or a perithecium.

Ostium (Ostia): In sponge, opening through which water is drawn into the body.

Otolith: Granule of calcium carbonate in vertebrate inner ear.

Outbreeding: Mating of genetically dissimilar, relatively unrelated individuals.

Ovarian follicle: Sac of cells which invests developing Oocyte of many Metazoa. Probably concerned with nourishment of growing oocyte and in vertebrates secretes oestrogen.

Ovary: The organ in which tile egg cells multiply and are nourished. Female gonad.

Oviduct (Zool): The tube by which eggs are conveyed from the ovary to the uterus or to the exterior.

Oviparous Egg-laying animals. Producing eggs that develop and hatch outside the mother's body.

Ovipositor: Organ at hind end of abdomen in female insects, through which eggs are laid.

Ovoviviparous: Embryos that develop to adult form within the mother's body while securing nourishment from the egg rather than directly from the mother's tissues. Many insects, snails, fishes, lizards and snakes are ovoviviparous.

Ovulation: Release of ovum from the ovary.

Ovule: Part of the female reproductive organs in seed plants. It consists of the nucellus, which contains the embryo sac, surrounded by the integuments.

Oxidase: Respiratory enzyme which brings about the oxidation of a substrate.

Oxidative decarboxylation: Oxidation by removal of carbon dioxide and hydrogen.

Oxidative phosph rylation: Process in which ATP molecule is formed from ADP and inorganic phosphate as a result of energy provided by oxidation Occurs in mitochondria

Oxidoreductases: A group of enzymes that catalyses oxidations and reductions.

Oxygenation: Temporary combination of oxygen with haemoglobin at the respiatory surface, oxygen is easily removed from it for cellular respiration.

Oxyntic cells: The acid (HCL) secreting cells of gastric gland; also called parietal cells.

P

Pachytene: Stage in prophase I of meiosis that is characterized by contraction of paired homologous chromosomes. At this point, each chromosome consists of a pair of chromatids and the two associated chromosomes are termed a tetrad.

Paedogenesis: Reproduction in larvae or other pre-adult forms, e.g. neotenic larvae like Axolotl larva of *Ambvstoma*.

Paedomorphic: Having the form of larva or other pre-adult stage.

Palaeoecology: Branch of biology dealing with prehistoric ecology as revealed by studying fossils and their artifacts, pollen samples and the mineral deposits.

Palaeontology: Study of extinct organisms, including their fossil remains, and impressions left by them; both extinct plants and animals.

Palaeozoic: The first and oldest era in which life became abundant, about 570 till 225 million years ago.

Palate: Roof of vertebrate mouth. In mammals and crocodiles roof of mouth is not homologus with that of other vertebrates; a new (false) palate has developed beneath original palate by bony shelves projecting inwards from bones of upper jaw.

Palisade tissue: Elongated, thin-walled cells containing chloroplasts, lying below upper epidermis of dorsiventral leaf.

Palmate: Lobes of a leaf spread from the same point, like the fingers of a hand.

Palmella: In various unicellular algae after division the daughter cells remain within the mucilaginous envelope of the parent cell and are thus rendered immobile. The cells may continue dividing giving a multicellular mass.

Palp: Tactile appendages on the head.

Palynology: Science dealing with the study of the morphology of pollen grains preserved in deposits, by scanning electron microscope, etc., to determine the nature of the flora.

Pancreas: A gland of gnathostome vertebrates situated in the mesentery near duodenum into which it discharges through hepato-pancreatic duct an alkaline mixture of digestive enzymes (trypsinogen, lipase, amylase, maltase, etc.). Also produces hormones insulin and glucagon.

Panicle: A type of inflorescence. A compound raceme formed by branching of peduncle, each branch bearing a raceme, e.g. oat.

Pantothenic acid: Vitamin of B group, forming a co-enzyme. Occurs in rice bran and plant and animal tissues.

Papain: A proteolytic enzyme in the unripe fruit and green leaves of papaya. Used as a meat tenderizer.

Papilla: Any nipple-like structure.

Pappus: Hairs developed on the calyx in Compositae. Act as a parachute and aid in wind-dispersal of fruits.

Parabiosis: Lateral fusion of embryos.

Paraphysis: Sterile filamentous structures occurring among reproductive structures in some algae, ascomycetes, basidiomycetes and mosses.

Parapodia: Paired, segmentally arranged, muscular lateral projections of body of polychaete worms, (e.g. *Nereis)*, bearing chaetae and sometimes other structures; locomotor in free-living forms.

Parasite : An organism that lives in/on another organism (host) and obtains from it food and shelter. It does not give any benefit to the host.

Parasympathetic (nervous system): A subdivision of the autonomic nervous system whose centres are located in the brain and lower portion of the spinal cord. Its ganglia are also paired like sympathetic system but these occur nearer to or within visceral organs. Two systems regulate the functioning of all internal organs. Its action is inhibitory, e.g., vagus nerve releases acetylcholine to slow down heartbeat.. All parasympathetic nerves release acetylcholine at synapse. Its action is antagonistic to that of the sympathetic division.

Parathormone: Hormone of parathyroid glands involved in calcium and phosphorus metabolism.

Parathyroid glands: Four small oval-shaped endocrine structures embedded in the thyroid gland. They produce a hormone that controls the blood-calcium level.

Paratonic: When a plant movement is induced by external stimulus.

Paratype: Any specimen other than the holotype cited with the original taxonomic description.

Parazoa: Grade of organization exemplified only by Porifera, which are multicellular animals, separately evolved from all others, e.g. sponges (Sycon, Spongilla).

Parenchyma (Bot): Thin-walled oval or spherical cells having intercellular spaces. Pith and cortex made of it. (Zool.). Mesechyme tissue of acoelomate animals.

Parietal cells: Gastric cells which secrete hydrochloric acid and intrinsic factor.

Paripinnate: A compound pinnate leaf which has an even number of leaflets.

Parotid: One of the three pairs of salivary glands.

Parotid glands: Glands in the head region of toad which produce irritating secretions.

Parthenocarpy: Development of fruit without fertilization. Such fruits, e.g. banana and pineapple, are seedless but otherwise normal in appearance. Parthenocarpy can also be induced in unfertilized flowers e.g. in tomato, by application of certain auxins.

Parthenogenesis: Development of ovum without fertilization. In many animals it may be induced artificially. In some plants (dendelion) and animals (aphids, rotifers, wasps. Russian lizard-Laceria sexicola) is of normal occurrence. Ova which develop in this way are usually diploid and all offspring is genetically identical with the parent.

Particulate inheritance: Inheritance of distinctive characteristics from both the parents in the progeny

Parturition: The act of giving birth to the foetus at the termination of pregnancy in viviparous animals.

Passage cell: Thin-walled cell in the endodermis of a root, found opposite the protoxylem element, through which water diffuses into the pericycle.

Passive absorption: The absorption of a substance by a cell according to laws of diffusion, i.e. along the concentration gradient, without expenditure of energy,

Passive immunity: Resistance of infection resulting from the direct transfer of antibodies or T cells from one individual (usually animals) to another.

Pasteurization: Method of partial sterilization of drinks in which heating at temperature well below boiling point destroys bacteria. Best known example is pasteurization of milk (heating for 30 minutes at 62°C), which kills tubercle and other harmful bacteria. Developed by Louis Pasteur.

Patella (Knee cap): A bone (sesmoid bone) over the front of the knee-joint in tendon of (extensor) muscles which straightens the hind leg. Present in most mammals, some birds and reptiles.

Pathogen: Disease-causing microorganism.

Pathology: Study of diseases or diseased tissue.

Peat: An accumulation of partly decomposed plant matter.

Pectinase: An enzyme destroying pectin of the middle lamella of cell walls. Used in protoplast fusion technique.

Pectoral: Region of body bearing forelimbs.

Pectoral girdle (Shoulder girdle): A bony or cartilaginous skeletal structure in the shoulder of vertebrate body to which forelimbs or fins are attached.

Pedicel: The stalk of a flower.

Pedigree: Family tree. A record of ancestry of an individual.

Pedipalp: one of a pair of the head appendages of chelicerate arthropods.

Pelagic: Animals and plants living in the open upper water.

Pelecypoda (Bivalvia): A class of mollusca. Shell of two lateral values hinged together dorsally. No head, jaws or radula. Foot hatchet shaped. Sexes separate, gills plate-like Includes clams, mussels, shipworms, oysters.

Pellagra: A deficiency disease due to the lack of nicotinic acid (niacin), a member of vitamin B complex.

Pellicle: Outer protective covering of many unicellular organisms, especially flagellated and ciliated protozoans. It is thin and flexible and is made up of protein. It maintains shape of the body.

Pelvic girdle (Hip girdle): A skeletal structure in the posterior region of the vertebrate body, to which hind-limbs or fins are attached.

Pelvis: (1) The pelvic girdle. (2) The lower part of the abdomen surrounded by pelvic girdle. (3) Pelvis of kidney—funnel-shaped expansion of ureter as it joins the concave side of kidney.

Pentadactyl limb: A limb having five digits; its bone components have a basic arragement characteristic of all tetrapod vertebrates.

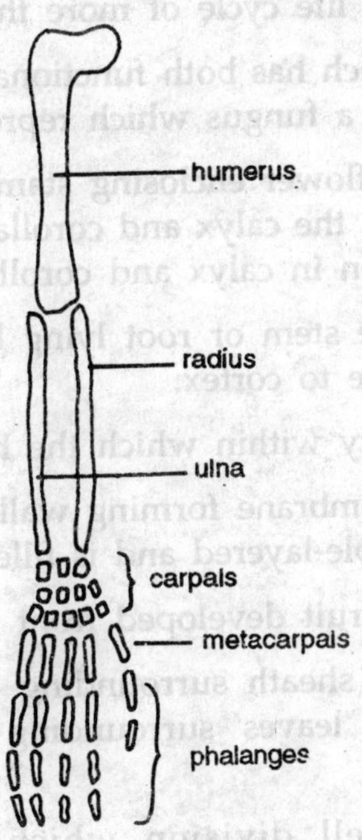

The pentadactyl limb

Pentose phosphate cycle: When oxygen is deficient, this cycle oxidises glucose-6-phosphate by oxidized NADP, resulting in the formation of pentose phosphate and the release of carbon dioxide.

Pepsin: Enzyme secreted in vertebrate stomach that splits proteins in acidic medium.

Pepsinogen: An inactive precursor of pepsin.

Peptidase: An enzyme that splits peptides, and in many cases proteins, by attacking certain peptide links.

Peptide: A short polypeptide chain (having—NH—CO bond).

Peptide bond: Bond between amino acids in proteins. Formed between amino and carboxyl groups of adjacent amino acids by elimination of a water molecule.

Peptone: Product of protein splitting. Consists of large fragments.

Perennation: Survival of plants from year to year by vegetative means (with the help of stored food).

Perennial: Plants having life cycle of more than two years.

Perfect: (I) A flower which has both functional anthers and ovules. (2) Stage in the life-cycle of a fungus which reproduces by sexual means.

Perianth: Outer part of flower enclosing stamens and carpels, usually consisting of two whorls, the calyx and corolla, or any one of them. In moncots no differentiation in calyx and corolla.

Periblem: A part of the stem or root lying between the dermatogen and plerome. It gives rise to cortex.

Pericardial cavity: Cavity within which the heart lies.

Pericardium: Serous membrane forming wall of pericardial cavity in the vertebrates. It is double-layered and is filled with pericardial fluid.

Pericarp: The wall of a fruit developed front the ovary-wall.

Perichaetium: A cuplike sheath surrounding the archegonia in some liverworts. The group of 'leaves' surrounding the sex-organs of some mosses.

Periclinal: Plane of cell division which is parallel with the circumference.

Pericycle: Outermost layer of stele in vascular plants. Present between endodermis and vascular bundles.

Periderm: Secondary protective tissue formed in secondarily-thickened stems and roots. Consists of the phellogen (cork cambium), phellem (cork), and phelloderm (secondary cortex).

Perigynous: A flower in which the receptacle develops into a concave structure, on which the sepals, petals and stamens are borne. Half superior.

Periosteum: Layer of connective tissue tightly investing vertebrate bones, to which muscles and tendons are attached, Contains active or (in adult) potential osteoblasts which in embryo and in adult after fracture are important in formation of bone.

Peripheral: To or toward the surface; away from the centre.

Peripheral nervous system: The test of nervous system when the central nervous system is excluded. It consists of cranial and spinal nerves.

Peristalsis: Rhythmic involu-ntary muscular contraction passing along tubular organs especially of the digestive tract and move the contents along.

Peristome: Fringe of elongated teeth around the mouth of capsule of moss.

Perithecium: Rounded or flask-shaped fruit-body of certain ascomycetes and lichens. They have an internal hymenium of asci and paraphrases, with an apical pore through which the ascospores are discharged

Peritoneal cavity: Abdominal cavity, Coelomic cavity of mammals posterior to diaphragm, containing liver, spleen, most of gut and other viscera that almost completely fill it.

Peritrichous: When flagella are distributed over the whole surface of the bacterial cell.

Permanent wilting: A state of wilted leaves or plants from which they cannot recover even if supplied with water.

Permease: An enzyme responsible for translocation of ion(s) across a membrane.

Permeability: Property of a membrane to allow the molecules to pass through it, Substances differ greatly in the case with winch they can pass through a given membrane.

Permian: Uppermost period of the Palaeozoic era from 280-235 million years ago, dominated by a few types of reptiles and gymnosperms.

Pernicious anaemia: *R.B.C.* fail to proliferate and mature due to deficiency of vitamin B_{12}.

Pest: An animal or plant causing considerable damage to crops, animals or possessions.

Pesticide: A chemical used to kill pests.

Petal: One of parts forming corolla of a flower. Often brightly coloured.

Petrification: Change of organic structures, such as a tree, into mineral structure.

pH: Negative logarithm of the hydrogen ion concentration (in moles/litre) of a solution. Provides a measure of acidity and alkalinity. pH less than 7 is acidic and more than 7 is alkaline.

Phaeatophyte: Plant dependent upon ground water as opposed to soil moisture.

Phaeophyceae (Phacophyta): Brown algae. It has brown pigment fucoxanthin. Thallus is multicellular and large.

Phagocyte: Any cell that engulfs foreign particles into its cytoplasm from its surroundings. Phagocytes are an important defence mechanism of most Metazoa against invading bacteria. In man and other mammals, polymorphs and macrophages are phagocytic.

Phagocytosis: Engulfing of solid particles by a cell.

Phalanges: Finger or toe digit bones. Each finger has one to five phalanges (more in whales) joined end-to-end in a row, the proximal of each row being jointed to a metacarpal bone.

Phanerpgams: An old term used for seed-bearing plants.

Phanerophyte: A woody tree or shrub with resting buds freely exposed on branches raised above the soil.

Pharynx: Throat, The common passage for food and air.

Phellem: (Cork) Outer zone of suberized cork cells formed by cork-cambium during secondary growth of stems or roots.

Phelloderm: Secondary cortex formed by cork-cambium on the inner side during secondary growth of stems or roots.

Phellogen: Cork cambium. A layer of lateral meristem lying in the cortex of a stern or root. It forms cork on its outer surface and phelloderm on inner.

Phenocopy: Phenotype of a given genotype changed by external condition to resemble the phenotype of a different genotype.

Phenology: Study of periodicity phenomena in plants, e.g., time of flowering in relation to climate.

Phenotype: The morphology of an organism produced by the reaction of a given genotype with the environment.

Pheromones: Chemical substances released into the surroundings by some orga-nisms. These influence the behaviour of other individuals of the same species. For example, sexual attractants of many insects.

Phenylketonurea: Genetically controlled disorder. It is characterized by the excertion of very large quantity of phenylketones in the urine and is associated with mental retardation.

Phoneme: One of the basic sounds of which the speech is made up.

Phosphatase: An enzyme which splits phosphate radicals from their organic compounds.

Phospholipid: A lipid derivative in *which* one fatty acid has been replaced by a phosphate group and one of several nitrogen-containing molecules.

Phosphorylation: Addition of a phosphate group to an organic molecule. For example, addition of phosphate to ADP to synthesize A T P.

Photic zone: The surface of an ocean or lake through which light penetrates and in which the phytoplankton flourish. Red and yellow wavelengths of light penetrate to about 54m while blue and violet light reach 200m.

Photomicrograph: A photo-graph obtained with a microscope

Photomorphosis: Change in the structure of a plant after exposure to strong fight.

Photonasty: Nondirectional paratonic movement due to light stimulus, e.g. the opening and closing of flowers.

Photophile: 'Light-loving' light receptive phase of circadian rhythm lasting about 12 hours.

Photopigment: Light-sensitive molecule for photic energy of certain wavelengths. Consists of an opsin bound to chromophore.

Photophosphorylation: (Photosynthetic phosphorylation). Process in which ATP molecule is formed from ADP and inorganic phosphate, using the light energy absorbed by chlorophyll.

Photoperiodism: The response of plants to the relative length of day and night. Flowering and many other plant response are controlled by photoperiod.

Photoreceptor: Receptor detecting light, e.g. eye.

Photorespiration: Oxidation of glycolate produced during glycolate metabolism of photosynthesis of many green plants. It uses oxygen and produces carbon dioxide. It wastes carbon and energy, using more ATP than it produces. It is estimated that in C_3 plants 40% of the potential yield of photosynthesis is lost through photorespiration. It occurs in peroxisomes and uses enzyme glycolate glyoxylate oxidase.

Photosynthesis: Synthesis of monosaccharides by green plants from water and carbon dioxide using energy absorbed by chlorophyll from sunlight. Oxygen is liberated during the process. Takes place within chloroplasts. In photophosphorylation (light reaction) light energy is converted into chemical energy in the form of energy, rich ATP. In the dark reaction, using the energy of ATP, carbon dioxide is reduced to glucose.

Photosynthetic number: The ratio between the number of grams of carbon dioxide absorbed per hour by a unit of leaf to the number of grams of chlorophyll which the unit contains.

Photosynthetic quotient or ratio: During photosynthesis, the ratio between the volume of carbon dioxide absorbed to the volume of oxygen set free over a fixed period of time.

Phototaxis: Paratonic movement (taxis) of locomotion of the organism in response to light.

Phototropism (Heliotropism): Paratonic growth movement of part of 1 plant in response to light, e.g. bending of stems of indoor plants towards a window; brought about by increased elongation of cells in growth region at tip of stem on shaded side.

Phragmoplast: In plant cells equatorial region of the spindle between the two groups of chromosomes separating from one another at anaphase of mitosis. Across it at telophase in equatorial plane, cell plate develops.

Phycobilins (Biliproteins). Red (phycoerythrin) and blue (phycocyanin) pigments present with chlorophyll in red and blue-green algae. Sec Phycoerythrin and Phycocyanin.

Phycocyanin: The blue pigment in blue-green algae.

Phycoerythrin: The red pigment of the Rhodophyta.

Phycology: The study of algae.

Phycomycetes: Lower fungi called algal fungi. Typically aseptate and multinucleate. Saprophytic or parasitic.

Phylloclade: Modified stem. The stem is flattened like a leaf and turns green in colour to carry on photosynthesis. Leaves either fall down or are modified into spines to check transpiration. For example, cactus.

Phyllode: Petiole modified into leaflike flattened green structure to make up for loss of leaves. An adaptation to check transpiration

Phyllotaxis: The arragement of leaves on a stem. There may be one, two, or several leaves at each node.

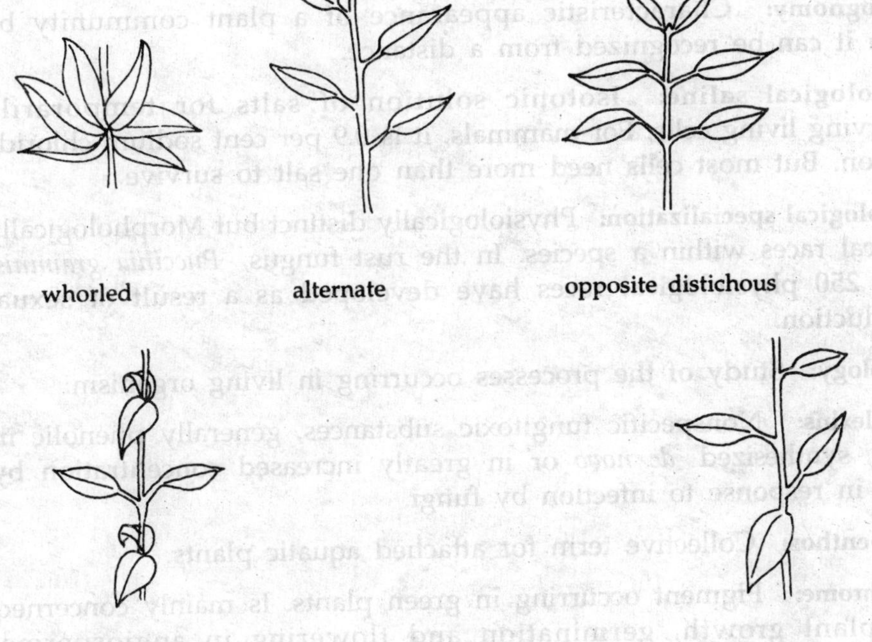

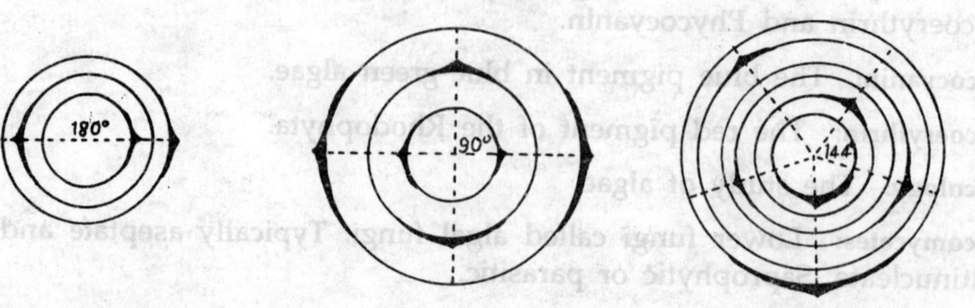

Phyllotaxis

Phyllotaxy: The arrangement of leaves on the stem whorled, opposite and alternate.

Phylogeny: Evolutionary history of a species.

Phylum: Major taxonomic category comprising one or more classes. One of the major kinds of group used in classifying animals, e.g. phylum Chordata. In plant classification, the term division is often used instead

Physiognomy: Characteristic appearance of a plant community by which it can be recognized from a distance.

Physiological saline: Isotonic solution of salts for temporarily preserving living cells, For mammals, it is 0.9 per cent sodium chloride Solution. But most cells need more than one salt to survive.

Physiological specialization: Physiologically distinct but Morphologically identical races within a species. In the rust fungus, *Puccinia graminis*, about 250 physiological races have developed as a result of sexual reproduction.

Physiology: Study of the processes occurring in living organism.

Phytoalexins: Nonspecific fungitoxic substances, generally phenolic in nature, synthesized *de novo* or in greatly increased concentration by plants in response to infection by fungi.

Phytobenthos: Collective term for attached aquatic plants.

Phytochrome: Pigment occurring in green plants. Is mainly concerned with plant growth, germination and flowering in angiosperms.

Phytochrome occurs in two forms. The form which absorbs red light (660 nm) is designated PR and the form which absorbs far-red light (740 nm) is termed PFR. The former is blue-green and latter is light-green in colour. PR stimulates flowering and PFR inhibits flowering. The two forms are interconvertible.

Phytotoxic: A chemical toxic to plants or plant parts.

Phytotron: Experimental buildings maintained at a variety of controlled conditions in which plants can be grown on a large scale for research.

Pia-arachnoid membranes: Delicate membranes enclosing vertebrate central nervous system; piamater immediately covering central nervous system, arachnoid outside that and in contact with outermost pia mater. Filled with cerebrospinal fluid.

Pia-mater: The connective tissue membrane that closely adheres to and covers brain and spinal cord.

Pileus: The cap of fruiting body of mushroom bearing hymenium (bearing gills) on the underside.

Pineal body (Epiphysis): Assumed to secrete a hormone melatonin, whose function is not conclusively known.

Pili (Fimbriae): Fine hair-like protein structures produced by the walls of certain bacteria, They are meant for attachment. Sex-pili participate in conjugation.

Piliferous layer: The part of a root epidermis which bears root-hair.

Pinea apparatus: Primitively consists of two outgrowths of roof of fore-brain lying within skull. Anterior is parietal organ forming pineal eye in some lizards and sphenodon (tuatara), a vestigial eye in lamprey, and absent in other vertebrates. Posterior forms pineal body or epiphysis

Pinna: A wing or fin; the projecting part of the external car in mammals.

Pinnate: Compound leaf with leaflets arranged in two ranks on opposite side of the rachis.

Pinocytosis: Bulk ingestion of liquid fluid by cell. Local invagination of plasma membrane completely surrounds minute drop of fluid which is thus engulfed by cytoplasm.

Pioneer community: The first plant community to become prominent on formerly bare ground e.g. lichens on a bare rock.

Pisces: Fishes. Aquatic and respire by gills. Skin covered by scales; cold-blooded. Heart two chambered. Divided into two major classes-chondrichthyes (cartilagenous fishes) and osteichthyes (bony-fishes).

Pistil: Each separate carpel of an apocarpous gynoecium. The gynoecium as a whole, whether it is apocarpous or syncarpous.

Pith: Central core of usually parenchymatous cells in those stems in which vascular tissue is in the form of a cylinder.

Pitressin: Extract of posterior lobe of pituitary, causing constriction of capillaries and arterioles when injected (which raises arterial pressure) and diminished urine formation.

Pits: Small, sharply defined areas in walls of plant cells that remain thin while rest of wall becomes thickened.

Pituitary body: Endocrine gland beneath floor of brain within skull of vertebrates. Secretes a number of hormones, all proteins or glyco-proteins. Anterior lobe produces 6 hormones; follicle-stimulating, lutenizing, lactogenic, ACIH, thyrotropic and growth hormone (somatotropic hormone, STH); its activities are influenced by hormones from the hypothalamus. Intermediate lobe produces intermedin, concerned with changing skin colour in fish, amphibia and reptiles. Posterior lobe secretes oxytocin and ADH.

Placenta (Bot): Part of ovary wall on which ovules are borne, (Zool) Organ consisting of embryonic and maternal tissues in close union, by which embryo of viviparous animals is nourished.

Placentalia (Eutheria Placental mammals): A sub-class of Mammalia, comprising the great majority of living mammals. The embryo develops in the maternal uterus, attached to the maternal tissues by placenta.

Placentation: Arrangement of placenta in an ovary. It may be axile, parietal, free central, marginal and basal.

Placoid scales: Very small, plate-like dermal exoskeletal scales.

Plagioclimax: A subclimax due to the action of biotic factors. Grazing, for example, prevents grassland from developing into climax.

Plagiogeotropism: Orientation of plant part by growth curvature in response to stimulus of gravity so that its axis makes an angle other than a right-angle with the line of gravitational force; e.g., branches of a main root which make an acute angle with the vertical.

Plantigrade: Walking on ventral surface (sole) of whole loot. e.g. man.

Planktdn: Free-floating animals (zooplankton) and plants (phytoplankton) living near the surface of a sea or lake, which float or drift almost passively. They are mostly very small; the smallest are nanoplankton e.g. diatoms.

Planosome: An odd chromosome resulting from, non disjunction of a pair during meiosis.

Plantation crops: Growing of crops like cotton, coffee, tobacco, etc.

Plant geography (Phytogeo-graphy): Study of ranges of plants over the earth and of the causes, present and past, underlying their characteristic distributions.

Plant sociology (phyto-sociology): Study of plant communities that make up vegetation, their origin, formation and composition structure.

Plaque: A clear area caused by a bacteriophage in a bacterial colony on an agar plate caused by localized destruction of bacterial cells.

Plasma: The fluid portion of blood or lymph, containing salts, proteins etc.

Plasma cells: Cells (leucocytes) derived from lymphocytes in the lymphoid tissue. They produce antibodies, important in defence against disease.

Plasmagene: Particle in cytoplasrn having the properties of a gene in that it is self-reproducing, shows interitance from cell to daughter cell, and affects the character of the cell bearing it; but unlike a gene it is not inherited in Mendelian fashion through the gametes.

Plasma proteins: The number of dissolved proteins present in vertebrate blood vessels.

Plasmid: The genetic element present and replicating autonomously in the cytoplasm of cells.

Plasmodesmata: Extremely fine cytoplasmic threads, one to few tenths of a micrometre wide, passing through walls of plant, cells and connecting cytoplasm of adjacent cells.

Plasmodium: (Bot.) Vegetative stage of slime-fungi (Myxomycophyta). Multinucleate, amoeboid mass of protoplasm bounded by plasma membrane, without a definite size or shape. (Zool.) Malarial parasite, a sporozoan, causing malaria.

Plasmolysis: The shrinkage of the protoplasm front the cell wall due to exosmosis.

Plasmon: Total complement of extra-chromosomal hereditary factors (plasmagenes) in a cell.

Plasmosome: Nucleolus not composed of chromatin.

Plastachron: Time interval between two of a series of periodic events, e.g. development of leaf primordia.

Plastids: Small, variously shaped bodies in cytoplasm of plant cells (excluding bacteria, blue-green algae, fungi, slime-fungi), one to many per cell in different plants. These are of three types-leucoplasts (colourless), chromoplasts (pigmented) and chloroplasts (green).

Plastogene: Plastid carrying a factor having properties of a gene but unlike a gene not inherited in Mendelian fashion. A plasmagene.

Platelets (Thrombocytes): Small flattened cells in vertebrate blood which help in clotting of blood by producing thrombokinase at the time of injury.

Platyhelminthes: Invertebrate phylum of flatworms—tapeworms and flukes.: Triploblastic and bilaterally symmetrical animals. Digestive tract absent or without anus. Coelom absent and body cavity filled with parenchyma.

Plectenchyma: A thick tissue formed by hyphae becoming twisted and fused together. Two types prosenchyma and pseud oparenchyma.

Pleistocene: The first Geological epoch of Quaternary of the Coenozoic era, from about two million years ago until the last glaciation ended about 10,000 years ago. During it, four major ice ages occurred. Succeeded by Recent epoch.

Plerome: The central core of tissue in the meristerm (stem or root apex) of vascular plants. It gives rise to the vascular cylinder and pith.

Pleura: Serous membrane covering the lung and lining the inner wail of the thorax.

Pleuron: Lateral plate on either side of a somite in arthropods

Plexus: A network of interlaced nerves or blood vessels.

Pliocene: The epoch of the Tertiary period of coenozoic era, about 71.5 million years ago. In the Pliocene, the hominids such as *Australopithecus and Homo,* which eventually led to man, became clearly distinguishable from apes.

Plumule: (Zool), Down feather of birds. Temporary feather of nestling bird, persisting, in some adult birds between contour leathers. No barbule. (Bot.). Apical part of an embryo in seed plants, which gives rise to the shoot.

Pneumatophore: (Bot) Special root branch of some vascular plants growing in water or in swamps, e.g. mangrove; grows erect, projecting into the air above for respiration. (Zool.). A bag-like medusoid form filled with secreted gas in some polymorphic hydrozoans for equilibrium, e.g. *Physalia.*

Pneumograph: An instrument for recording respiratory movements.

Pod: A simple, dry dehiscent fruit developing from a single carpel, having a single loculus containing one (rarely) to many seeds. It dehisces by splitting along both ventral and dorsal sutures. Characteristic of Leguminosae.

Poikilothermal (Cold-blooded): Animals with fluctuating body temperature, which follows that of the surroundings. Characteristic of ail animals except birds and mammals.

Polar: Bacterial flagella arising at one or both ends of cell

Polarity: Intrinsic antero-posterior orientation in an organism. Characteristic of most organisms, unicellular and multicellular, and of many cells within organisms.

Polar body (Polocyte): Small cell formed during formation of oocyte. Contains one of the nuclei derived from first or second meiotic division. But no cytoplasm. Soon dies.

Polarized: Having two electric poles—one negative and one positive. Pollen. Microspores (male gametes) of seed plants. Unicellular.

Pollex: Thumb of pentadactyl forelimb.

Pollination: Transfer of pollen from anther to stigma.

Pollinium: A coherent mass of pollen grains held together by a sticky substance, and transported as a whole during *pollination. For* example, orchids.

Pollution: Deterioration in the quality of environment by impurities,

Polyadelphous: Stamens joined by their filaments only to form several bundles.

Polyandry: Mating of one female with many males.

Polycyclic: A stele which has two or more concentric rings of vascular tissue, as in ferns.

Polycythemia: A condition of excess red blood cells in blood.

Polydactylous: Having more than the normal number of digits.

Polydipsia: Consumption of increased amount of water. Excessive thirst.

Polyembryony: (Bot.) The formation of more than one embryo within a seed of a flowering plant. The extra eggs may be derived from vegetative nucellar cells, sister mother-cells, sister spores or sister nuclei within one embryo-sac. (Zool.). Asexual formation of more than one embryo per zygote by fission at some early stage of development, e.g. armadillo (*Dasypus*) always has identical quadruplets from a single ovum. Monozygotic twins are the simplest form of polyembryony in man.

Polyestrous cycle: Repro-ductive cycle which occurs several times a year (e.g. pig).

Polygamy: Mating with more than one partner of opposite sex.

Polygenes (Multiple factors): Genes with individually a very small effect on phenotypic differences, when several genes control the same character to produce the conti-nuous quantitative variation such as height, weight and skin colour.

Polygyny: Mating of one male with several females.

Polymodal: Neuron that responds to the stimuli of more than one modality.

Polymorphism: Occurrence of several distinct phenotypes in a species e.g. queen, drone and worker bees in the same habitat of a species.

Polymorphonuclear granu-locyte (Polymorph, Granu-locyte): Type of white blood cells which have a nucleus of irregular shape.

These form between 6575 per cent of white blood cells in man.

Polynucleotide: Molecule consisting of several nucleotide units DNA and RNA are polynucleotides.

Polypeptide: Molecule consisting of fewer than 100 amino acids, linked together in a single chain.

Polypetalous: When petals are free from each other in a flower.

Polyphyletic: A group of species classified together is polyphyletic when some of its members have had quite distinct evolutionary histories, not being descended from a common ancestor which was also a member of the group. Derived from more than one ancestral type.

Polyplanetism: When zoospores have resting and motile phases alternating with each other.

Polyploid: When the organism has chromosome number three or more times the haploid number, designated as 3x, 4x, 6x, 8x, etc.

Polysaccharide: Carbohydrate that is made up of 3 or more monosaccharides linked together. For example, starch, cellulose.

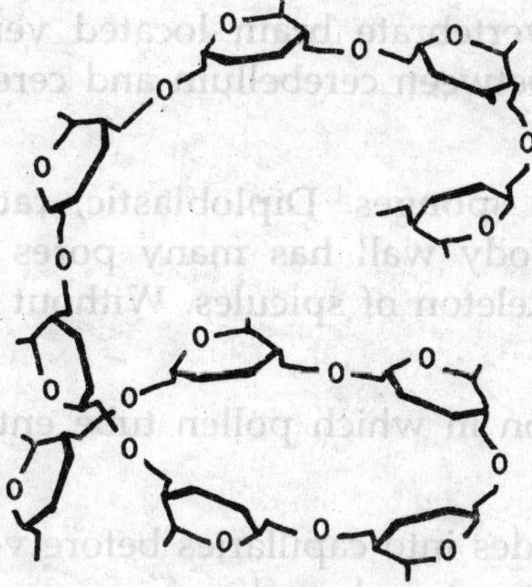

Polysaccharide

Polysepalous: Flowers with sepals free from one another.

Polysiphonous: When an algal thallus consists of a central row of elongated cells surrounded by one or more layers of peripheral cells as in *Polysiphonia*.

Polysomes (Polyribosome): Cluster of ribosomes that is formed during protein synthesis. Ribosome is the cell organelle which is responsible for protein synthesis

Polysomic: When one chromosome is represented more than two times in a diploid organism.

Polyspermy: Penetration of many sperms into one ovum during fertilization. Occurs normally in very yolky eggs (shark, bird), but only one sperm nucleus fuses with the egg nucleus, the rest disintegrate. Polyspermy can also be induced artificially.

Polystelic: Having more than one stele.

Polytene: Giant chromosome consisting of numerous parallel identical chromatids due to repeated duplication without division, as in the case of salivary gland cells of dipterous insects, e.g. *Drosophila*.

Polythetic: System of classification in which membership of taxon is based on possession of large number of common characters.

Polytopic: A taxon occurring in two or more separate areas.

Polyuria: The excretion of large volume of dilute urine.

Pome: A false fruit, which develops from the receptacle of the flower, and not from ovary.

Pomology: The study of cultivated fruits and fruit trees.

Pons: A region of vertebrate brain located ventrally that serves as connecting pathway between cerebellum and cerebrum. Floor of fourth ventricle.

Porifera: Phylum of sponges. Diploblastic, radially symmetrical or without symmetry. Body wall has many pores and canals for water circulation. Has exoskeleton of spicules. Without mouth, digestive tract or other organs.

Porogamy: Fertilization in which pollen tube enters the ovule through micropyle.

Portal vein: That divides into capillaries before vein reaching the heart. Has a set of capillaries at each end.

Posterior: Toward the hinder end, away from the head. Opposite of anterior.

Postganglionic neuron: Neuron of autonomic nervous system whose cell body lies in a ganglion and whose axon junctions with gland ' smooth or cardiac muscle. It conducts impulse away from ganglion towards periphery.

Potential: Voltage difference between two points.

Prairie: Grasslands of cold climates.

Precipitin reaction: In vitro reaction between solutions of antigen and antibody, with the formation of a precipitate. Important technique in diagnosis of human pathogens and plant pathogenic viruses and the basis for serological identification of microorganisms.

Predator: An animal that preys upon other animals for its food.

Preformation: A preformed miniature individual already contained in the gamete (male or female), homunculus according to scientists of 18th century. Embryonic development supposedly consisting merely of enlargement and manifestation of this structure.

Pregnancy: Condition of actually bearing an embryo or foetus within the uterus of fallopian tubes in mammals

Prehensile: Adapted for grasping or holding.

Premolars: Crushing teeth (bicuspid) of mammals, having usually more than one root and a pattern of ridges and *projections of* biting surface. Replaceable along with incisors and canines in man once unlike molars.

Presentation time: The minimum time for which an organism is exposed to a stimulus before there is a perceptible response.

Presumptive: (Zool.). Embryological term applied to tissues before their differentiation, meaning 'becoming in the course of normal development'.

Primary meristem: Meristem that has persisted from origin in embryo. For example, apical meristem, cambium.

Primary oocyte: Germ cell of the female. It undergoes first meiotic division to form a secondary oocyte and a polar body.

Primary spermatocyte: Male germ cell. It undergoes meiosis to form two secondary spermatocytes.

Primary succession: A succession of plant communities starting on a bare uninhabited soil or water area.

Primate: Any member of the order Primates of placental mammals, including man, apes and monkeys. A primarily arboreal order, rather primitive in structure. Climb by grasping, nails commonly present instead of claws; and big toe and thumb usually well-developed and opposable to other digits. Well-developed eyesight, often with binocular vision, relatively large brain size.

Primine: Outer integument of an ovule.

Primitive: Not specialized; the early state in evolutionary history of a species~

Primitive streak: Longitudinal thickening in disc-line early embryo of birds or mammals during gastrulation. Produced by accumulation of mesoderm as this material moves from its superficial position into the interior of the embryo.

Primordial germ cells: Large cells in vertebrates which appear early in embryonic life. These migrate to developing gonads and there give rise

Primordium: Earliest development stage in the formation of an organ or body part.

Proamnion: The area Pellucida immediately anterior and lateral to the head process of early chick embryo, consisting only of ectoderm and endoderm.

Proboscis: Tubular extension at the anterior end of an animal; used in feeding.

Procambium: Part of apical meristem which gives rise to the vascular tissues.

Procoelous: Concave in front, as the centrum of some vertebrae.

Proctodaeum: An intucking of ectoderm of embryo with endoderm at the posterior end of gut, forming anus or cloacal opening (vent).

Procumbent: Stem which lies prostrate on the ground.

Productivity: Rate of dry matter production by photosynthesis in an ecosystem.

Proembryo: Early stage in the development of an embryo of seed plants

Progametangium: A short side-branch from which a gametangium is developed in the members of Mucorales.

Progesterone: Steroid hormone secreted by the corpus luteum in the ovary after ovulation. It initiates preparation of the uterus for implantation of the ovum, development of the placenta and the development of mammary glands in preparation of lactation.

Proglottid: One of the segments of a tapeworm.

Prokaryotes (Procaryotes): The primitive organisms lacking true nucleus. Comprise bacteria and blue-green algae (cyano-bacteria) Nuclear membrane is absent and 'chromosome' is represented by DNA alone. Histones are absent. Ail membrane-bound organelles like mitochondria, golgi bodies, lysosomes, etc., and endoplasmic reticulum are absent.

Prolaction: Leuteotropic hormone. A lactogenic hormone produced by the anterior pituitary gland. In mammals stimulates the mammary glands for milk production after estrogen and progesterone priming.

Promeristem: Extreme tip of apical meristem consisting of actively dividing cells which have net yet begun to show differentiation.

Prometaphase: The stage between the dissolution of the nuclear membrane and the aggregation of the chromosomes on metaphase plate.

Promycelium (Epibasidium): A short germ-tube, formed by some fungal spores, on which other spores of different types develop.

Pronation: Rotation of the forearm so that the hand is twisted through 90 degrees in either direction in relationship to the elbow. Pronation in man occurs when palm faces downwards or backwards and the radius and ulna are crossed.

Propagule: A small reproductive branch which becomes detached front the parent and grows into a new plant.

Prophase: A stage during cell division when chromosomes become distinct and nuclear membrane disappears.

Proplastids: Immature, colourless plastids occurring in cells of meristem. Consist of a double membrane enclosing granular stroma. Multiply by division. Give rise in mature cells to leucoplasts or chromoplasts.

Proprioceptor: Receptor which detects position and movement; balancing organs of internal car.

Prosenchyma: Type of fungal hyphae in which component hyphae which are loosely woven can be recognized.

Prostaglandin: Any of a number of 20-carbon organic acids that are synthesized in the body from unsaturated fatty acids and are responsible for a variety of metabolic activities. They act as chemical messengers.

Prostate gland: A gland in male mammals surrounding the urethra, near the bladder. Releases a fluid containing enzymes and an antiagglutinating factor, that contributes to the production of semen.

Protandrous: The male gametes developing before the female gametes. The anthers ripening before the stigma of the same flower become receptive.

Protease: An enzyme that hydrolyses peptide bonds whether of proteins or peptides. It decomposes proteins into amino acids.

Protein: Large polymer comprising one or more linear sequences of amino acid sub-unit joined by peptide bond.

Proteinase: Enzyme that hydrolyses peptide bonds in proteins.

Proteolysis: Hydrolysis of proteins into component amino acids. Enzymes that catalyse this reaction are proteases or proteolytic enzymes.

Proteolytic enzyme: An enzyme breaking down proteins.

Prothallus: Independent gametophyte plant of Pteridophyta. Small, green, parenchymatous thallus bearing antheridia and archegonia.

Prothrombin: Inactive precursor of enzyme thrombin. Produced by liver and present in plasma.

Protista: Group (Kingdom) originally proposed by Haeckel for ail unicellular, eukaryotic organisms like algae, slime molds, protozoa and fungi.

Protochordata: Chordata without vertebral column.

Protocorm: A tuber-like structure formed in the early stages of the development of club-mosses and some other plants that appear to live in close association with fungi when young.

Protogamy: Union of gametes without the fusion of their nuclei.

Protogynous: The development of the female gamete before the male gamete. The stigma of a flower being receptive before the anthers or the same flower discharge their pollen.

Protonema: Branched, multicellular, filamentous structure produced on germination of spore in Bryophyta, from which new plants develop as buds.

Protonephridium: Excretory organ in Platyhelminthes, Nemertinea, Rotifera and sortie trochophore larvae. Consists of one or more flame cells at the inner end of tubes which open to the exterior.

Protophloem: The first phloem formed from procambial strands with poorly formed sieve tubes and no companion cells.

Protoplasm: The living substance of a cell.

Protoplast: The protoplasm of a plant cell within cell wall or cell membrane.

Protostele: Simplest type of stele found in lower pteridophytes; consists of a central core of xylem elements surrounded by phloem.

Protoxylem: First xylem element which differentiates from procambium. Appear on the inner side next to pith.

Protozoa: Phylum of unicellular animals. Eucaryotic. Without tissues or organs. Found in water or in moist places. May be free-living (amoeba) or parasites (malarial parasite).

Proventriculus: Anterior part of stomach of birds where enzymes are secreted; the posterior part being the gizzard or ventriculus where grinding of grains and digestion occurs.

Proximal: Near the place of origin or attachment.

Psammophyte: A plant which occurs only in sand.

Pseudo-alleles: Alleles in the sense that they affect the same process in organism and involve homologous cistrons as judged by the cis-trans affect, but nevertheless, because they originated as mutations in different parts of the cistron, show recombination by crossing-over within the cistron.

Pseudocoelomate: Animals in which spaces between body wall and internal organs are not fined by cells of mesodermal origin.

Pseudogamy: Development of ovum into a new individual as a result of stimulation by a male gamete whose nucleus however, does not fuse with that of the ovum and contributes nothing to hereditary constitution of embryo. Occurs in some nematodes and in some higher plants.

Pseudomixis: Fusion between two vegetative cells or between cells which are not differentiated as gametes.

Pseudomycelium: When the cells are loosely united in chains as a result of repeated budding in yeast.

Pseudoparenchyma: A mass of loosely interwoven filaments which looks like parenchyma, in fungi.

Pseudopodium: Temporary protoplasmic protrusion of the cell, associated with flowing movements and nutrition in rhizopods and amoebocytes for feeding and locomotion.

Psychrophilic: Microorganisms which thrive at temperatures below 20°C.

Pteridophyta: A division of plants, comprising ferns, club mosses, etc. Mainly terrestrial. Have true (vascular) stem, leaves and roots. Well marked alternation of generation. For example, ferns.

Pterodactyla (Pterosauria): Fossil order of flying-reptiles. Lived during Jurassic and Cretaceous. Membraneous wings supported mainly by greatly elongated fourth finger, and also by rest of arm.

Ptyalin: A starch-splitting enzyme of saliva; salivary amylase of man.

Ptyxis: The way in which young leaves are roll in a bud.

Puberty: Age at which an individual begins to become sexually mature between 1215 in girls and 13,16 in boys. Secondary sexual characters appear at ibis age; for example facial hair in boys and breast development in girls.

Pubis (Pubic bone): One of a pair of bones forming the anterior ventral portion of the tetrapod pelvic girdle. They are sometimes joined to form the pubic symphysis as in most mammals, ostrich and some reptiles.

Pubic symphysis: Joint formed mid-ventrally by the fusion between pubic bones of two halves of pelvic girdle. In most mammals, ostrich and reptiles.

Pulmonary: Pertaining to the lungs.

Pulmonary artery: A paired artery that carries deoxygenated blood from the right ventricle of the heart to the lungs.

Pulmonary edema: Accumulation of fluid in the lung interstitium and air sacs

Pulmonary circulation: Flow of blood from the heart to the lungs and back.

Pulmonary vein: Carries oxygenated blood from lung to heart.

Pulp cavity: Internal cavity of a vertebrate tooth. Opens usually by a narrow channel m the tissues in which tooth is embedded; containing conne-ctive tissue, nerves and blood-vessels, with odontoblasts lining the dentine wall of the cavity.

Pulse: (Zool.) A rhythmical dilation of the artery caused by the systolic output of the heart. (Bot.). Plants of pea family, rich in proteins.:

Pulvinus: Localized enlarge-ments of the base of leaf-stalk and or of base of leaflets in certain plants. Concerned with the movement of leaves or leaflets in response to stimuli.

Pupa: Stage usually dormant, between the larva and the adult of insects having complete metamorphosis. Pupa of a butterfly is called chrysalis.

Puparium: The case formed by the hardened last larval skin in which the pupa is formed.

Pupil: The central aperture of the iris through which light passes to, reach retina.

Pure line: The formation of homozygous individuals that are identical to each other and continue to breed true.

Purine bases: Adenine and guanine. Double-ring base sub-units of nucleotides..

Purkinje cells: Pear-shaped nerve cells, with extensive dendrites in the cerebellar cortex and forming its only axial output.

Pustule: A mass of fungal spores and hyphae bearing them.

Putrefaction: Bacterial decom-position of protein-containing substrate (largely anaerobic) with formation of foul-smelling amines rather than ammonia.

Pycnidiospore: A spore formed in a pycnidium in fungi and lichen.

Pycnidium: It is a minute flask-shaped fruit body in fungi, with apical ostiole lined internally with conidiophores.

Q

Quadrat: Sampling unit chosen at random for study of the composition of vegetation in a selected area, may be a square or a circle.

Quadriplegia: Paralysis of the forelimbs of the body.

Quadrulus: One of the three groups of longer cilia (other two are dorsal and ventral peninculus) fining the buccal cavity part of the oral groove of *Paramecium;* beating causes ingestion of food and anticlockwise rotational and spiral movements of animal during locomotion.

Qualitative Inheritance: Inheritance in which expression of a particular character differs sharply amongst individuals of a species (discontinuous variation). For example, sex.

Quantasomes: Photosynthetic units. Present as very minute particles on the membranes of the thylakoids of grana within the chloroplasts.

Quantitative inheritance: Inheritance in which expression of the character concerned differs only in degree amongst members of a species, characteristically showing every gradation from one extreme to the other (continuous variation) with a predominance of intermediate types. For examples, stature in man.

Quarantine: All operations associated with the prevention of importation of unwanted organisms into a territory.

Quartan lever (malaria): Caused by *Plasmodium malarae* recurrence every 72 hours; less fatal; predominant in tropics and subtropics.

Quaternary: Current period of the Coenozoic from 1.5 million years ago, comprising pleistocene and recent.

Queen: Fertile female of eusocial insects.

Quills: Large feathers on the wings and tail in birds; with a long shaft and broad vane made of barbs and barbules.

Quiscent zone: The terminal part in apical meristem of the root apex, where mitotic divisions are absent. They become meristematic if other part of meristem is injured.

QRS Complex: Component of the electrocardiogram corresponding to the depolarization of ventricles.

Racemose inflorescence: An inflorescence in which the growing point of the axis continues to develop and produce lateral branches; i.e. show monopodial growth.

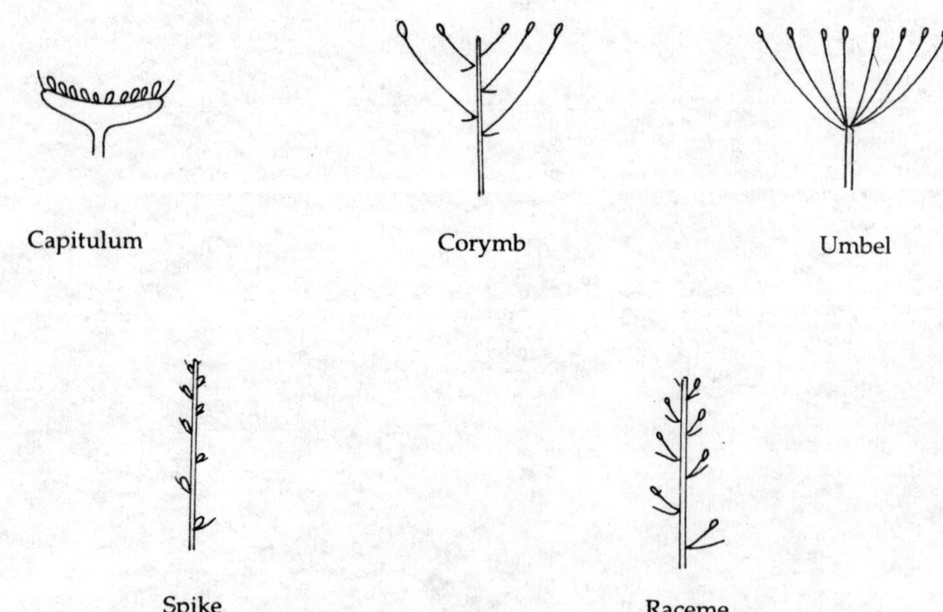

Types of racemose inflorescence

Rachis: Petiole of a compound leaf.

Rachitis: Disturbance in bone formation; rickets.

Rad: A unit of dosage of any radiation equal to 100 ergs of energy for one gram of mass of material irradiated.

Radial symmetry: Similar parts arranged around a common central axis, as in a starfish.

Radiata: Includes the coelenterates, ctenophores and many sponges, because of their primary and secondary radial symmetry.

Radical: When leaves arise from the base of the plant at ground-level.

Radicle: Part of embryo which develops into roots.

Radiobiology: Science dealing with the effect of radiation on living organisms.

Radius: Anterior, inner and shorter of the two bones (other is ulna) of forearm of tetrapod forelimb.

Radula: Tongue' and 'rasping organ' in the buccal cavity of molluscs; a horny strip with numerous rows of replaceable teeth (seven in each row) on its surface.

Ram: Male sheep.

Ramenta: Fibrous remains of leaf bases that cover rhizome.

Ramus: Branch or outgrowth of a structure.

Ramet: Independent individual of a clone.

Raphe: 1. Vertical marking on seed coat developed from an anatropous ovule; the portion of the funicle is fused with the integument to form it. 2. A slit found in the valve of diatoms.

Raphides: Needle-like crystals of calcium oxalate occurring in bundles in some plants.

Ray: The non-vascular tissue in a stele. That between the primary vascular bundles in the interfascicular ray and that develops in the secondary vascular tissue, by division of the cambium, is a vascular ray.

Reaction time: Time that elapses between stimulation of a whole organism and the production of a detectable response.

Recapitulation theory (Bio-genetic law): Theory proposed by Haeckel that embryological development of an organism summarizes the evolutionary history of the species, i.e. ontogeny repeats phylogeny. Reflection of ancestral history (phylogeny) i.e. the characters of the ancestral adult forms in the developmental stages (ontogeny) of an organism.

Recent (Holocene): The present epoch of coenozoic era in the Geological Time Scale from the end of the last glaciation, about 10,000 years ago, to the present,

Receptacle (Thalamus): The apex of a reproductive branch on which the reproductive organs are borne.

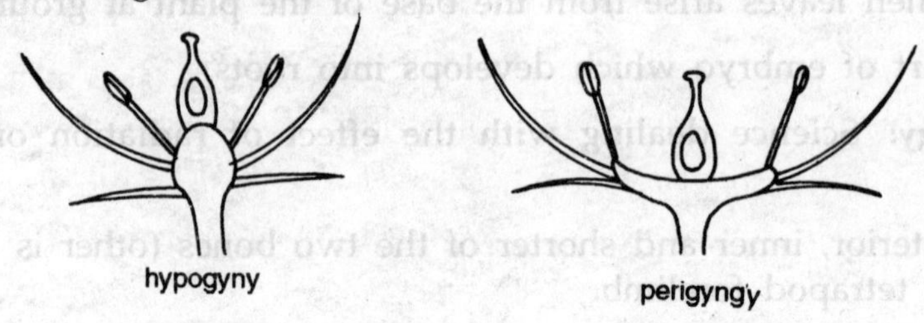

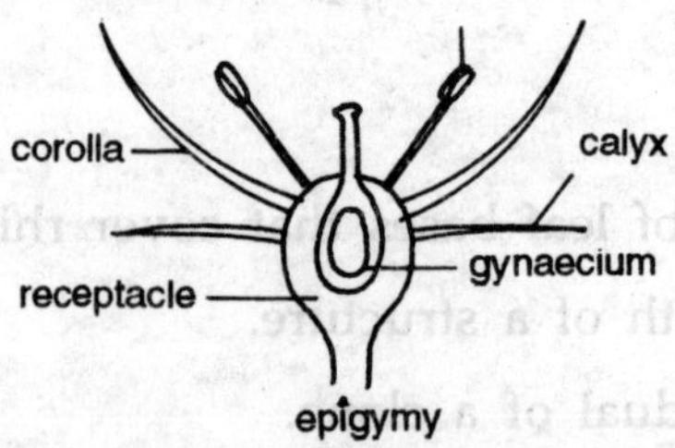

The various arragements of the floral parts on the receptacle

Receptor (Sense organ): That organ of an animal which in cooperation with the nervous system, detects what goes on, i.e., receives stimuli from outside or inside the animal.

Recessive: An allele which faits to express in heterozygous condition; expressed in the phenotype only in the homozygous condition.

Reciprocal cross: A cross which tests whether the inheritance of′ a particular character is affected by the sex of the parent. The cross is thus made both ways, i.e. the character under consideration is carried by the female in one cross and by the male in the second cross.

Recombinant DNA: DNA which is experimentally formed by joining portions of two DNA molecules earlier fragmented by restriction enzyme.

Recombinant gametes: Gametes with new combinations on chromosomes which are formed as a result of crossing over.

Recon: Smallest genetic unit that can undergo crossing over.

Rectum: Terminal part of intestine opening to exterior by anus, or cloacal opening.

Red blood cells or corpuscles (Erythrocytes): Red cells of vertebrate blood, containing haemoglobin. Carry nearly all the oxygen contained in the blood. In mammals have no nuclei, except in embryo. Formed in myeloid tissue of bone marrow and perish continuously in enormous numbers in man roughly five million per cubic millimetre of blood. Life is 120 days in man and 100 days in frog.

Redia: Larval stage in the life-cycle of flukes. It is produced by the sporocyst larva and in turn it produces many cercariae, This stage is passed in snail.

Reductase: Enzyme which promotes a reduction reaction.

Reflex: An involuntary response to a stimulus.

Reflex arc: A sequence of sensory, association and motor neurons which conduct the nerve impulses for a given reflex.

Refractory period: Brief interval following the response of a neuron or muscle fibre during which it is incapable of a second response.

Regeneration: Regaining of tissues or organs lost or healing of the wounds. In some Platyhelminthes minute fragment can regenerate the whole animal; in mammals limited to wound-healing, regrowth of peripheral nerve fibres, liver and compensatory hypertrophy. Very common in plants, occurring e.g., in higher plants by growth of dormant buds, formation of secondary meristems and production of adventitious buds, roots.

Regulator gene: Gene in bacteria which produces a repressor substance that binds to the operator gene.

Releaser: Stimulus that initiates instinctive behaviour.

REM (Roentgen equivalent man): The amount of absorbed radiation that will cause as much damage in human tissue as one rad of X-rays.

Renal: Pertaining to the kidney.

Renal portal system: On its way to the heart, the blood in fishes and Amphibia is brought from capillaries of posterior end of body to kidneys by a pair of renal portal veins which break up into capillaries in the kidney. It is a system of veins having capillaries at both ends.

Renin: A hormone produced by the kidney which causes an increase in blood pressure by its role in the production of angiotensin.

Rennet: Extract of the fourth stomach of the calf, containing rennin.

Rennin: A milk-clotting enzyme in the stomach of human infants and calves, converting soluble milk protein caseinogen into casein which forms insoluble calcium-casein compound.

Replacing bone: Cartilage-bone.

Replication: Synthesis of new DNA molecules during cell division.

Repressor: Substance produced of regulator gene which binds to the operator gene.

Repressor gene: Portion of DNA that contains the genetic instructions for the synthesis of repressor gene.

Reproduction: The maintenance of a species from generation to generation.

Reptilia: A class of phylum chordata. Epidermis covered with cornified scales. Heart incompletely four-chambered. Cold-blooded. Fertilization internal. Breathing by lungs, for example, lizard.

Residual volume: Volume of air remaining in the lungs after maximum expiration.

Respiration: Intake of oxygen, its transportation to the cells and its utilization to release energy and Co_2 and back transport of this CO_2 releasing it ultimately in the surroundings, is known as respiration. It involves breathing.

Respiratory pigment: Substance present in the blood to combine reversibly with 0, to transport it. Various such pigments are haemoglobin (iron pigment) dissolved in the plasma of earthworm and i some arthropods and in the RBCs of all vertebrates; haemocrythrin is also an iron pigment in the blood of some arthropods and molluscs; haemocyanin (copper pigment) imparting bluish-green tinge to the blood of most molluscs and few arthropods.

Respiratory quotient (R.Q.): Ratio of the volume of carbon dioxide expired to the volume of oxygen consumed during the same time. For carbohydrates it is 1, for fats, 0.7 and for proteins 0.8.

Response: Any change or activity resulting front a change in either the internal environment of organism.

Resting membrane potential: Voltage difference between the outside of a cell in the absence of excitatory or inhibitory potential.

Restriction enzyme: Bacterial enzyme that splits DNA into many fragments, acting at different loci in its two strands and at points containing a particular sequence of nucleotides.

Rete cords: Strands of epithelial cells, containing many primordial germ cells which connect with the seminiferous tubules and later become vase efferentia in birds.

Reticulate: Pertaining to net-like structure.

Reticulocyte: Immature mammalian red blond cell which has lost its nucleus but shows some staining due to RNA.

Reticulo-endothelial system: Phagocytic macrophages, of vertebrates in ,contact with blood (in bone-marrow, spleen, liver) or lymph (in lymph nodes); they free these fluids front foreign particles, e.g. bacteria.

Retina: Innermost of the three layers of vertebrate eye, except in front (in region of cellar body); it is sensitive to light.

Retinal: Derivative of vitamin A. It forms chromophore component of a photopigment.

Reversion: The reappearance of ancestral traits which have been in abeyance for one or more generations.

Rf value: In paper chromatography, distance travelled by the solute divided by the distance travelled by the solvent front. Characteristic of a particular molecule.

Rhizoids: Short, hair-like outgrowths in lower plants; help in nutrition and attachment to the ground.

Rhizome: Elongated, underground horizontal stem.

Rh factor: Group of RBC plasma membrane antigens which may (Rh;) of may not be (Rh) present. Important for blood transfusion.

Rhizopoda (Sarcodina): A class of protozoans that have pseudopodia for food capture and locomotion, e.g. Amoeba.

Ribose: Pentose sugar present in ribonucleic acid.

Rhodophyta (Redalgae): The group of algae characterized by their red colour. They contain pigments, phycoerythrin and phycocyanin.

Rhodopsin: A pigment in the rods of retina, sensitive to dim light. Breaks in the light and reformed in the darkness.

Riboflavin (Lactoilavin, Vitamin B_2): One of the water-soluble vitamin of B-group. Forms part of co-enzymes concerned in cellular oxidation.

Ribonucleic acid (RNA): Nucleic acid of single strand of nucleotides made up of ribose sugar base (adenine, guanine, cytosine, uracil) and phos-phate group. Responsible for protein synthesis by decoding genetic information. Exists in three forms—mRNA, rRNA, tRNA.

Ribosomal RNA (rRNA): Type of RNA which is synthesized in nucleus but moves to the cytoplasm and forms part of the ribosomes.

Rickets: Disease in which new bone matrix is not adequately calcified due to deficiency of vitamin.

Rigor mortis: The stiffening or hardening of muscles soon after death. Complete after 12 hours.

Ringer (Ringers fluid): Physiological saline containing sodium, potassium and calcium chlorides. Used for temporarily maintaining cells or organs alive in *vitro*.

Rod cell: One of the visual cells of the retina.

Rodent: Any member of the order Rodentia, e.g. rat and squirrel. Gnawing mammals, with a pair of large continually growing chiesel-like incisors in upper and lower jaws.

Root cap: Mass of cells covering the apical meristem of root.

Root hair: Unicellular, root-epidermal outgrowth. Present near root-tip. Absorbing organs.

Root nodules: Small swellings on the roots of leguminous plants. These contain nitrogen-fixing bacteria.

Rotation of crops: Sowing of cereal and legume crops in successive seasons.

Rotifera: Wheel-animalcules. Phylum of minute Pseudo-coelomates. Metazoa which swim and feed by ciliated band, the 'wheel'

Roundworms: Belong to class nematoda of phylum nemathelminthes (aschelminthes). Unsegmented, cylindrical and elongated worms without cilia.

Rudimentary: Incompletely developed or having no function.

Rumen: Storage compartment, first and the largest of the four chambers (rumen, reticulum, omassum, abomassum) of a ruminant's, complicated stomach. Newly eaten but unchewed food is passed and from which it (the 'cud.) is subsequently returned to mouth for chewing. Some digestion of food (especially cellulose) and much synthesis of B vitamins occur by bacterial action.

Ruminant: Some mammals of the other Artiodactyle, having rumen as one of the four chambers of the stomach, e.g. deer, giraffe, sheep, goat etc. No upper incisor teeth.

S

Sacral vertebra: Vertebra of tetrapods, which articulates hip region with ilia of hip-girdle.

Sacromere: Repeating structural unit of a muscle myofibril. The region of a myofibril extending between two adjacent Z lines.

Sacrum: Fused sacral vertebrae in the lower back region that are attached to the ilia and give support to the pelvic girdle of tetrapods. Five fused sacral vertebrae form, sacrum in man,

Sagittal: In the plane extending longitudinally and dorso-ventrally in the mid-line. Divides bilaterally symmetrical animal into two similar (right and left) halves.

Saliva: Secretion of salivary glands a water solution of salts and proteins, including mucus and ptyalin.

Salivary glands: Three pairs of glands in the mouth which secrete saliva.

Salivary gland chromosomes: Giant chromosomes occurring in salivary glands of dipterous insects (including *Drosophila).* Chromosomes microscopically visible unlike in normal resting nuclei.

Sapwood: Outer region of xylem of older tree trunks.

Saprophyte: Heterotrophic plant that secures its food by the extracellular digestion of organic matter.

Saprozoic: Organisms that absorb dead organic matter as solution through body surface.

Sarcolemma: Extensible sheath of a muscle fibre enclosing a contractile substance.

Sarcoplasm: Interfibrillar protoplasm of muscle fibres.

Sarcoplasmic reticulum: Endoplasmic reticulum of muscle fibres.

Savanna: A tropical grassland consisting of tall, coarse grasses and scattered small trees.

Scalariform: Ladder-like.

Scala tympani: Fluid-filled inner car compartment which receives sound waves from basilar membrane.

Scansorial: Pertainrng to or adapted for climbing.

Scape: Leafless flowering stem arising from ground level.

Scapula: A large, flat, triangular bone forming the dorsal, portion of the pectoral girdle.

Scavanger: An animal that feeds on dead plants and animals or on animal faeces.

Schistosoma (Bilharzia): Trematoda: Platyhelminthes. Human blood parastie. Digenic like liverfluke, causes schistosomiasis.

Schwann cell: Kind of cells around every nerve-fibre of vertebrate peripheral nervous system. Has Schwann nucleus.

Scion: Twig of one plant that is grafted to the stock of another

Sclerenchyma: Spindle-shaped cells (fibres) with thick walls. Dead at maturity. Give mechanical support to the plant.

Sclerotic: Fibrous, thick and firm outermost coat of eyeball in the vertebrates, continuous with cornea in front of eye.

Scrotum: Bag-like pouch of the skin of pelvic region in most mammals, containing testis. Keeping them cooler than body temperature.

Scurvy: A pathological condition due to lack of vitamin C. Gums bleed.

Sebaceous gland: Epidermal skin gland of mammals, embedded in dermis, nearly always opening into a hair-follicle, secreting fatty substance (sebum).

Secondary sexual characters: Those characters which distinguish one sex from the other, excluding sex organs

Secondary meristem: Region of active cell division which arises from permanent tissues. For example, cork cambium.

Secondary xylem: Xylem formed as a result of secondary growth.

Secretion: The passage of material produced by a cell from the inside to the outside of its plasma membrane.

Secretin: A hormone secreted by the Brunner's glands in the wall of duodenum and jejunum under stimulus of acidic food coming from stomach. Stimulates pancreas to release $NaHCO_3$ to act as buffer.

Seed: Product of fertilized ovule. Comprises an embryo enclosed by protective coat. Reserve food may be present in endosperm or in cotyledon.

Segment: A part that is marked off from others; any of the several divisions of a body or an appendage.

Segmentation: Synonym for cleavage.

Segmentation cavity: Cavity of blastula. Syn, blastocoel, subgerminal cavity.

Segregation: Separation of two members of any pair of allelomorphs possessed by an individual during gamete formation; the two allelomorphs having previously been brought together in the individual, one from each of its parents and having in no way blended or altered each other while associated together in the individual. Mendel's first law asserts that allelomorphs segregate. Segregation is the result of the relatively unchanging nature of genes.

Seismonasty (Bot.): Response to a non-directional shock stimulus.

Self-fertilization: Fusion of male and female gametes from the same individual.

Self-pollination: Transfer of pollen from anther to stigma of the same flower or to stigma of another flower of the same plant.

Semen: Sperms together with secretions of various accessory glands.

Semicircular canals: For equi-librium. Three looped canals that form part of the labyrinth of vertebrate inner ear.

Sinus: An anatomical cavity, space, or channel. Examples are the nasal sinuses in the skull.

Skeleton: The hardened framework of an animal body serving for support and protecting soft parts; it may be external or internal.

Skeletal muscle (Striated, Striped or Voluntary muscle): Muscles that are attached to the bones. Each muscle is made of many muscle fibres bound together with connective, tissue and surrounded by a sheath, reach fibre is long and narrow with tapering ends, an outer Membrane (sarcolemma) and many nuclei, Its cytoplasm (sarcoplasm) contains many longitudinal myofibrils.

Skin: the outer layer of the body of an animal. In vertebrates it protects the animal from excessive loss of water, from the entry of disease-causing organisms, from damage by ultravilet radiation, and from mechanical injury. It contains numerous nerve endings and therfore also acts as a peripheral sense organ.

Sleep: A normal recurrent state of reduced responsiveness to exteral stimuli in vertebrates. Sleep in man is characterised by typical brain wave patterns, recorded as electroencephalograms, which demonstrate the existence of different phases of sleep. Other mammalian species show comparable EEG patterns, but in lower vertebrates the characteristic signs of sleep may vary and a formal definition of the state becomes more difficult.

Sliding-filament mechanism: Sliding of actin filaments on myosin filaments with the utilization of ATP. It is responsible for contraction of muscles.

Sliding growth: A pattern of plant growth seen, for example, in many epidermal cells where, in order to accommodate growth by adjoining cells, expanding cell walls side along each other.

Sludge: Waste-water solids suspended in water.

Smog: Smoke along with fog. It is formed in industrial cities under cold conditions.

Smooth muscles (Involuntary or Unstriated muscles): Spindle shaped cells without transverse striations. Uninucleated. Involuntary. Found in gut, diaphragm etc.

Sodium: An eement essential in animal tissues, and often found in plats although it is believed not to be essential in the latter. It is found in bones, and is the most abundant ion in the blood and cell fluids, being extremely important in maintaing the osmatic balance of animal tissues.

Sol: A quasi-liquid state of a colloidal system in which the liquid phase is continuous and the solid phase is dispersed.

Solitary: Living alone; not in colonies or groups.

Soma: Cell body of a nerve cell.

Somatic: Pertaining to the body or body cells other than germ cells.

Somatic hybridization: Hybridization by the fusion of protoplasts of somatic cells (and not of sex cells).

Somatotrophic hormone: A secretion of adenohypophysis which accelerates growth of organism.

Somite: One of the blocks of mesoderm that develop in a longitudinal series on either side of the notochord in vertebrate embryos. A serial segment or homologous part of the body.

Somitic mesoderm: Mesoderm on dorsal side of vertebrate embryo. Becomes segmented into a series of blocks, called the somites.

Sorus: A group of sporangia borne on a sporophyll in ferns. It is covered by indusium.

Spadix: A type of spike inflorescence with thick fleshy axis, bearing small, usually unisexual flowers.

Spathe: A large coloured bract around a flower or inflorescence to attract for Pollination.

Spawning: Act of expelling mass of eggs from the uteri of amphibians.

Special creation: The hypothesis for origin of organisms. It proposes that every species was separately created by the action of a supernatural force.

Species: A group of interbreeding organisms which are reproductively isolated from other similar groups.

Spermatheca: An organ in female or hermaphrodite animal; receives and stores sperms.

Spermatium: Non-motile male sex-cell present in red algae and in fungi, in some members of the Ascomycetes and Basidiomycetes.

Spermatocyte: A diploid cell, within the seminiferous tubules of the testis, that develops during spermatogenesis. A primary spermatocyte resulting from a spermatogonium through first meiotic division produces two haploid secondary spermatocytes which through second meiotic division produce two spermatids each. Thus, one primary spermatocyte forms four spermatids, which later become spermatozoa.

Spermatogenesis: Production of spermatozoa within the testis. Primordial germ cells (2n) of the germinal epithelium lining the seminiferous tubules multiply by mitosis and form spermatogonia. They give rise to huge numbers of spermatozoa by meiosis continuously until the age of about 70.

Spermatogonium: Diploid reproductive cell formed through mitosis of the diploid primordial germ cells in the germinal epithelium that lines the seminiferous tubules of testis. It undergoes multiplication and growth to give rise to spermatocytes.

Spermatophyta (Seed plants): Seed-bearing plants, subdivided into Anglospermae and Gymnospermae.

Spermatozoon (Sperm) (Zool): Small, motile, usually flagellated male sex-cell (non-flagellated in *Ascaris*).

Sphincter: Circular muscle surrounding and closing an opening, e.g., pyloric sphinter and sphincter of the iris.

Sphygmomanometer: An instrument comprising an inflatable cuff and a pressure gauge for determining the blood pressure.

Spicules: Small, hard, calcareous or siliceous structure forming exoskeleton of Porifera.

Spike: A type of racemose, inflorescence characterized by sessile flowers borne on an elongated axis.

Spinal cord: Posterior part of the vertebrate central nervous system within the backbone. Single, hollow, dorsal, tubular, having a fine central canal. Grey matter inner to white matter.

Spindle: Body formed within a cell at mitosis or meiosis, taking part in the distribution of chromatids to the two daughter nuclei. Appears at metaphase, and chromosomes become arranged at its equator.

Spinnerets: Single in silkworm caterpillars to spin cocoon of silk, secreted in paired glands. In spiders, four to six elevations at the end of abdomen on to surface of which ducts of spinning glands open.

Spiracle: External openings of insect trachea for breathing; 10 pairs. Dorsally situated, reduced evolutionary remnant first gill slit in many fish.

Spiral cleavage: Due to cyclosis in cleaving cells, mitotic spindle gets twisted to right or left, thus resulting into spiral or oblique cleavage, e.g. nematodes, annelids, and molluscs.

Spiral valve: A valve in the truncus arteriosus of amphibian heart, guiding flow of different types of blood in the three aortic arches.

Spirillum: Cork screw-shaped bacteria with flagella at one or more poles.

Spirochaetes: Elongated spirally twisted unicellular bacteria having a flexible wall.

Spirometer: An instrument used for determining the volume of air inspired.

Spleen: A non-dispensible blood storage organ, manufacturing RBCs in foetus and lymphocytes in adult destroying dead RBCs. Mass of lymphoid tissue in mesentery of stomach or intestine of gnathostomes; but unlike lymph nodes as interposed in blood.

Spontaneous generation (Abiogenesis): Spontaneous origin of life from non-living things, Italian biologist, Francesco Redi in 17th century, first discarded this theory. But until the rise of bacteriology, initiated by Pasteur, spontaneous generation was widely believed to be for microorganisms.

Spongy parenchyma: Leaf mesophyll paremhyma with large intercellular spaces.

Sporocarp: A hard spore-containing structure found in water ferns (e.g. *Marsilea*).

Sporangiophore: Hypha bearing one or more sporangia in fungi.

Sporangium: Plant organ within which asexual spores are produced,

Spore mother cell: Diploid cell which by meiosis gives rise to four haploid spores.

Sporocyst: A larval stage in the life-cycle of flukes. It is produced by miracidium larva and it produces redia larva.

Sporogenous: Spore-producing.

Sporogonium: The sporophytes of bryophytes, always attached to the gametophyte.

Sporogony The division of a zygote or oocyst into spores.

Sporophore: The aerial spore-producing body of certain fungi, e.g., the mushrooms.

Sporophyte: Spore producing, diploid generation of plants giving rise asexually to haploid spores. It is the dominant generation in pteridophytes, gymnosperms and angiosperms while in bryophytes it is parasitic on the gametophyte.

Squamosal: Membrane bone of the skull, which in mammals takes over from quadrate, the articulation with lower jaw.

Stamen: Male reproductive organ in flowering plants. It is equivalent to the microsporophyll of the gymnosperms and heterosporous pteridophytes. Produces pollen grains.

Statocyst: Organ of equilibrium consisting of a vesicle containing granules of lime, sand etc., (statoliths) which stimulate sensory cells as the animal moves. Present in some Crustacea, flatworms, cnidarians, molluscs and echinoderms, etc.

Steady state: The state in which the inflow of energy and materials in an ecosystem is just sufficient to maintain biomass at a relatively constant level.

Stele: The cylinder of vascular tissue of vascular plant consists of xylem, phloem, pericycle, pith, medullary rays.

Sterile: (1) Incapable of reproducing sexually. (2) Free front micro-organisms. Sternum. (1) Breast bone of terrestrial vertebrates in the middle of the ventral side of the chest to which the ventral ends of most of the ribs are attached At its anterior end attached to shoulder girdle. (2) In insects, cuticle on ventral side of each segment, often forming a thickened plate.

Steroid: Any member of a group of compounds having a basic ring structure.

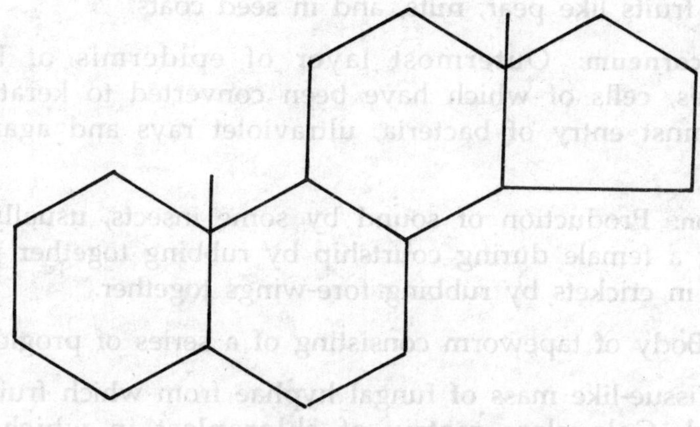

Steroid ring structure

Sterols: Lipid compounds, including cholesterol and its derivative steroid hormones, bile acids and certain vitamins. They are fat soluble. Their molecules contain four interconnected rings made up of seventeen carbon atoms.

Stigma: Terminal portion of style which receives pollen grains.

Stimulus: Any change in the environment of an organism which is intense enough to produce a change in the activities of the living material.

Stipule: Small leaflike structure on either side of leaf stalk in many plants.

Stock: Rooted stem part on which is grafted a part (scion) of another plant.

Stolon: Subaerial modification of stem. Grows horizontally and roots at nodes.

Stoma: Minute openings on leaf epidermis which allow exchange of' gases.

Stomodaeum: Foregut lined with ectodermis, an intucking of ecetoderm meeting endoderm of anterior part of gut, forming mouth.

Stratosphere: Atmospheric shell above troposphere and below mesosphere.

Stratified: A series of layers, one above another.

Stone cell (Sclereid): These are rounded or elongated sclerenchyma cells; found in fruits like pear, nuts, and in seed coats.

Stratum corneum: Outermost layer of epidermis of land-living vertebrates, cells of which have been converted to keratin. Protects body against entry of bacteria, ultraviolet rays and against loss of water.

Stridulation: Production of sound by some insects, usually males for attracting a female during courtship by rubbing together parts of the body, as in crickets by rubbing fore-wings together.

Strobila: Body of tapeworm consisting of a series of proglottides.

Stroma: Tissue-like mass of fungal hyphae from which fruit-bodies are produced. Colourless matrix of chloroplast in which grana are embedded. (3) Intercellular material of connective tissue component of an animal organ.

Structural gene: In protein synthesis, genes controlled by the operator are called structural genes.

Style: Prolongation of style that supports stigma.

T

Tannin: One of a group of complex organic chemicals commonly found in leaves, unripe fruits, and the bark of trees. Their function is uncertain though the unpleasant taste may discourage grazing animals.

Tapetum: A reflecting layer containing crystals of guanine, in the choroid of the eye of many nocturnal vertebrates. It reflects light back onto the retina, thus improving vision and causing the eyes to shine in the dark.

Tarsal (tarsal bone): One of the bones that form the ankle in terrestrial vertebrates.

Tarsus: The ankle (or corresponding part of the hindlimb) in terrestrial vertebrates, consisting or a number of small bones *(tarsals).* The number of tarsal bones varies with the Species: humans, for example, have seven.

Taste bud: A small sense organ in most vertebrates, speciali-zed for the detection of taste. In terrestrial animals taste buds are concentrated on the upper surface of the tongue. They are sensitive to four types of taste: sweet, salt, bitter, or sour. The taste bud transmits information about a particular type of taste to the brain via nerve fibres. The four types of taste bud show distinct distribution patterns on the surface of the human tongue.

Taxis (tactic movement): The movement of a cell (e.g. a gamete) or a microorganism, in response to an external stimulus. Certain micro-organism have a light-sensitive region that enables them to move towards or away front high light intensities (positive and negative *phototaxis* respectively). Many bacteria move in response to chemical stimuli *(chemo*taxis); a specific example *is aerotaxis, in* which atmospheric oxygen is the stimulus. Tactic movements are restricted to cells that possess cilia, flagella, or some other means of locomotion. The term is usually not applied to the movements of higher animals.

Taxon: Any named taxonomic group of any rank in the hierarchical classification of organisms. Thus the taxa Papilionidae, Lepidoptera, Insecta, and Arthropoda are named examples of a family, order, class, and phylum, respectively.

Taxonomy: The study of the theory, practice, and rules of classification of living and extinct organisms, The naming, description, and classification of a given organism draws on evidence front a number of fields, Classical taxonomy is based on morphology and anatomy. *Biochemical taxonomy* studies similarities in the structure of certain proteins and nucleic acids. *Cytoiaxonomy com*pares the size, shape, and number of chromosomes of different organisms. *Numerical taxonomy* uses mathematical procedures to assess similarities and differences and establish taxonomic groups.

Teleostei: The major subclass of the Osteichthyes (bony fish), containing about 20 000 species. Teleosts have colonized an extensive variety of habitats and show great diversity of form. The group includes the eel, seahorse, plaice, and salmon. They have been the dominant fish since the Cretaceous period.

Telophase: The fourth stage of cell division. In mitosis the chromatids that separated from each other at anaphase collect at the poles of the spindle. A nuclear membrane forms around each group, producing two daughter nuclei with the same number and kind of chromosomes as the original cell nucleus. In the first telophase of meiosis, complete chromosomes from the pairs that separated at first anaphase form the daughter nuclei. The number of chromosomes in these nuclei. The number of chromosomes in these nuclei. The number of chromosomes in these nuclei is therefore half the number in the original one.

Tendon: A thick strand or sheet of tissue that attaches a muscle to a bone. Tendons consist of collagen fibres and are therefore inelastic: they ensure that the force exerted by muscular contraction is transmitted to the relevant part of the body to be moved.

Tendril: A slender branched or unbranched structure found in many climbing plants. It may be a modified stem, leaf, leaflet, or petiole, Tendrils respond to contact with solid objects by twining around them. The cells that touch the object lose water and decrease in volume in comparison to the outer cells, thus causing the tendril to curve.

Tera-: Symbol T. A prefix used in the metric system to denote one million times.

Territory: A fixed area that an animal or group of animals defends against intrusion from others of its species. Outside the territory (which may contain food sources, hiding places, and nesting sites) others are not threatened. Many mammals indicate their territory boundaries with scent markings, while birds sing territorial songs that repel would be intruders. Animals in neighbouring territories normally: respect each other's boundaries, which reduces overt aggression.

Tertiary: The older geological period, of the Cenozoic era (*compare* Quaternary). It began about 65 million years ago, following the Cretaceous period, and extended to the beginning of the Quaternary, about 2 million years ago. It is subdivided into the Palaeocene, Eocene, Oligo-cene, Miocene, and Pliocene epochs in ascending order. The Tertiary period was characterized by the rise of the modern mammals and the development of shrubs, grasses, and other flowering plants.

Testa (seed coat): The lignified or fibrous protective covering of a seed that develops from the integuments of the ovule after fertilization.

Testis (testicle): The reproductive organ in male animals in which spermatozoa arc produced. In vertebrates there arc two testes; as well as sperm, they produce steroid hormones. In most animals the testes are within the body cavity but in mammals, although they develop within the body near the kidneys, they come to hang outside the body cavity in a scrotum. Most of the vertebrate testis is made up of a mass of seminiferous tubules, in which the sperms develop.

Tetrapod: A vertebrate animal with four limbs. Tetrapods include amphibians, reptiles, birds, and mammals. The skeleton of the limbs of all tetrapods is based on the same five-digit pattern.

Thalamus: (in anatomy) Part of lie vertebrate forebrain that lies above the hypothalamus. It relays sensory information to the cerebral cortex and is also concerned with the translation of impulses into conscious sensations.

Thallophyta: A former division of the plant kingdom containing relatively simple plants, i.e. those with no leaves, stems, or roots. It included the algae, bacteria, fungi and lichens.

Thallus: A relatively undifferentiated plant body with no true roots, stems, leaves, or vascular system. It is found in the algae, lichens, and bryophytes and in the gametophyte generation of pteridophytes.

Thermography: A medical technique that makes use of the infrared radiation front the human skin to detect an area of elevated skin temperature that could be associated with an underlying cancer. The heat radiated from the body varies according to the local blood flow, thus an area of poor circulation produces less radiation. A tumour, on the other hand, has an abnormally increased blood supply and is revealed on the *thermogram* (or *thermograph*) as a 'hot spot'. The techr..que is used particularly in mammogaphy, the examination of the infrared radiation emitted by human breasts in order to detect breast cancer.

Themoluminescence: Luminescence produced in a solid when its temperature is raised. It arises when free charge carriers, trapped in a solid as a result of exposure to ionizing radiation, unite and emit photons of light. The process is made use of in *thermoluminescent dating,* which assumes that the number of charge carriers trapped in a sample of pottery is related to the length of time that has elapsed since the pottery was fired. By comparing the luminescence produced by heating a piece of pottery of unknown age with the luimnescence produced by heating similar materials of known age, a fairly accurate estimate of the age of an object can be made.

Thigmiotropism (haptotropism): The growth of an aerial plant organ in response to localized physical contact. For example, when a tendril of sweet pea touches a supporting structure, it curves in the direction of the support and coils aroυnd it.

Thin-layer chromatography: A technique for the analysis of liquid mixtures using chromatography. The stationary, phase is a thin layer of an absorbing solid (e.g. alumina) prepared by spreading a slurry of the solid on a plate and drying it in an oven. A spot of the mixture to be analysed is placed near one edge and the plate is stood upright in a solvent. The solvent rises through the layer by capillary action carrying the components up the plate at different rates (depending on the extent to which they are absorbed by the solid). After a given time, the plate is dried and the location of spots noted. It is possible to identify constituents of the mixture by the distance moved in a given time. The technique needs careful control of the thickness of the layer and of the temperature.

Thoracic duct: The main collecting vessel of the lymphatic system, running longitudinally in front of the backbone. The thoracic duct drains its lymph into the superior vena cava.

Thoracic vertebrae: The vertebrae of the upper back, which articulate with the ribs. They lie between the cervical vertebrae and the lumbar vertebrate and are distinguished by a number of articulating facets for attachment of the ribs. In man there are 12 thoracic vertebrae.

Thorax: The anterior region of the body trunk of animals. In vertebrates it contains the heart: and lungs within the rib cage. It is particularly well-defined in mammals, being separated front the abdomen by the diaphragm. In insects the thorax is divided into an anterior *prothorax*, a middle *mesothorax*, and a posterior *metathorax*, each of which hears a pair of legs; the hindmost two segments also both carry a pair of wings. In other arthropods, especially crustaceans and arachnids, the thorax is fused with the head to form a *cephalothorax*.

Thorn: A hard side stem with a sharp point at the tip, replacing the growing point. In some plants the development of thorns and subsequent suppression of the growing points may be a response to dry conditions. Examples are the thorns of gorse and hawthorn.

Thread cell (nematoblast; cnidoblast): A specialized cell found only in the ectoderm of the Coelenterata. It contains a *nematocyst*, a fluid-filled sac within which lies a long hollow coiled thread. When a small sensory projection on the surface of the thread cell is touched, e.g. by prey, the thread is shot out and adheres to the prey, coils round it, or injects poison into it.

Threshold: (in physiology) The minimum intensity of a stimulus that is necessary to indicate a response.

Thrombin: An enzyme that catalyses the conversion of fibrinogen to fibrin.

Thymine: A pyrimidine derivative and one of the major component bases of nucleotides and the nucleic acid DNA.

Thymus: An organ, present only in vertebrates, that is concerned with development of lymphoid tissue and hence the antibody-producing white blood cells (lymphocytes) and the immune response. In mammals it is a bilobed organ in the region of the lower neck, above and in front of the heart. It shrinks in size after sexual maturity and is therefore believed to function only during early life.

Thyroid gland: A bilobed endocrine gland in vertebrates, situated in the base of the neck. It secretes two iodine-containing hormones, *thyroxine* and *triiodolyrosine*, which control the rate of all metabolic processes in the body and influence physical development. Growth and activity of the thyroid is controlled by a hormone, *thyrotrophin* (or *thyroid-stimulating hormone)*, secreted by the anterior pituitary gland.

Tibia: The larger of the two bones of the lower hindlimb of terrestrial vertebrates (*compare* fibula). It articulates with the lemur at the knee and with the tarsus at the ankle. The tibia is the major load-bearing bone of the lower leg.

Time-lapse photography: A form of cine photography used to record a slow process, such as plant growth, A series of single exposures of the object is made on cine film at predetermined regular intervals. The film produced is then projected at normal speeds and the process appears to be taking place at a most unusual rate.

Tissue: A collection of similar cells organized to carry out one or more particular functions, For example, in animals nervous tissue i specialized to perceive and transmit stimuli. An organ, such as a lung or kidney, contains many different types of tissues.

Tissue culture: The growth of the tissues of living organisms outside the body in a suitable culture medium. Culture (or nutrient) media contain a mixture of nutrients either in solid form (e.g. in agar) or in liquid form (e.g. in physiological saline). Tissue culture has proved to be invaluable for gaining information about factors that control the growth and differentiation of cells.

Titre: The number of infectious virus particles present in a suspension. A measure of the amount of antibody present in a sample of serum, given by the highest dilution of the sample that results in the formation of visible clumps with the appropriate antigen.

Tobacco mosaic virus (TMV): A rigid rod-shaped RNA-containing virus that causes, distortion and blistering of leaves in a wide range of plants, especially the tobacco plant. It is transmitted by insects when they feed on plant tissue. TMV was the first virus to be discovered.

Tollen's reagent: A reagent used in testing for aldehydes. It is made by adding sodium hydroxide to silver nitrate to give silver(I) oxide, which is dissolved in aqueous ammonia. The sample is warmed with the reagent in a test tube. Aldehydes reduce the complex Ag^{+} ion to metallic silver, forming a bright silver mirror on the inside of the tube.

Tomography: The use of X-rays to photograph a selected plane of a human body with other planes eliminated. The CAT (*computerized axial tomography*) scanner is a ring-shaped X-ray machine that rotates through 180° around the horizontal patient, making numerous X-ray measurements every few degrees. The vast amount of information acquired is built into a three-dimensional image of the tissues under examination by the scanners own computer. The patient is exposed to a dose of X-rays only seine 20% of that used in a normal diagnostic X-ray.

Tone (tonus): The state of sustained tension in muscles that is necessary for the maintenance of posture. In a tome muscle contraction, only a certain proportion of the muscle fibres arc contracting at any given time; the rest are relaxed and recovering for subsequent contractions. The fibres involved in tone contract more slowly than the fast fibres used for rapid responses by the same muscle. The proportions of slow and fast fibres depends on the function of the muscle.

Tongue: A muscular organ of vertebrates that in most species is attached to the floor of the mouth. It plays an important role in manipulating food during chewing and swallowing and in terrestrial species it bears numerous taste buds on its upper surface. In some advanced vertebrates the tongue is used in the articulation of sounds, particularly in human speech.

Tonsil: A mass of lymphoid tissue, several of which are situated at the back of the mouth and throat in higher vertebrates, In humans there are the *palatine* tonsils at the back of the mouth, lingual tonsils below the tongue, and *pharyngeal tonsils* (or *adenoids)* in the pharynx. They are concerried with the production of lymphocytes and therefore with defence against infection.

Tooth: Any of the bard structures in vertebrates that are used principally for biting and chewing food but also for attack, grooming, and other functions. In fish and amphibians the teeth occur all over the palate, but in higher vertebrates they are concentrated on the jaws. They evolved in cartilaginous fish as modified placoid scales, and ibis is reflected in their structure: a body of bony dentine with a central pulp, cavity and an outer covering of enamel on the exposed surface *(crown).* The portion of the tooth embedded in the jawbone is the root. In mammals there are four different types of teeth, specialized for different functions. Their number varies with the species.

Toxin: A poison produced by a living organism, especially a bacterium, *An endotoxin is* released only when the bacterial cell dies or disintegrates An *exotoxin is* secreted by a bacterial cell into the surrounding medium. In the body a toxin acts as an antigen, producing an immune response.

Trachea: The windpipe in air-breathing vertebrates: a tube that conducts air from the throat to the bronchi. It is strengthened with incomplete rings of cartilage. An air channel in insects and most other terrestrial arthropods. Tracheae occur as ingrowths of the body wall. They open to the exterior by *spiracles* and branch into finer channels (*tracheoles*) that terminate in the tissues. Pumping movements of the abdominal muscles cause air to be drawn into and out of the tracheae,

Tracheid: A type of cell occurring within the xylem of conifers, ferns, and related plants. Tracheids are elongated and their walls are usually extensively thickened by deposits of lignin. Water flows from one tracheid to an other through unthickened regions (pits) in the cell walls.

Tracheophyta: A group (division) of the plant kingdom containing an plants possessing organized vascular tissue. This classification is an alternative to other classifications, which divide vascular plants into the Pteridophyta and Spermatophyta. The Tracheophyta comprises the subdivisions Psilopsida (mostly extinct), Lycopsida, Sphenopsida, and Pteropsida.

Transcription: The process in living cells in which the genetic information of DNA is transferred to a molecule of messenger RNA (mRNA) as the first step in protein synthesis. Transcription takes place in the cell nucleus or nuclear region. An enzyme (transcriptase) proceeds along the DNA strand and assembles the nucleotides necessary to form a complementary strand of mRNA.

Transect: A straight line across an expanse of ground along which ecological measurements are taken, continuously or at regular intervals. Thus an ecologist wishing to study the numbers and types of organism at different distances above the low-tide line might sample at five-metre intervals along a number of transects perpendicular to the shore.

Translation: The process in living cells in which the genetic information encoded in messenger RNA (mRNA) in the form of a sequence of base triplets (codons) is translated into a sequence of amino acids in a polypeptide chain during protein synthesis. Translation takes place on ribosomes in the cell cytoplasm. The ribosomes move along the mRNA 'reading' each codon in turn. Molecules of transfer RNA (tRNA), each bearing a particular amino acid, art brought to their correct positions along the mRNA molecule: base pairing occurs; between the bases of the codons and the complementary base triplets of tRNA. In this way amino acids are assembled in the correct sequence to form the polypeptide chain.

Transpiration: The loss of water vapour by plants to the atmosphere, It occurs mainly from, the leaves through pores (stomate) whose primary function is gas exchange. The water is replaced by a continuous column of water moving upwards from the roots within the xylem vessels.

Trematoda: A class of parasitic flatworms comprising the flukes, such as *Fasciola* (liver fluke). Flukes have suckers and books to anchor themselves to the host and their body surface is covered by a protective cuticle. The whole life cycle may either occur within one hest or require one or more intermediate hosts to, transmit the infective eggs or larvae *Fasciola hepatica*, for example, undergoes larval development in a land mail (the intermediate host) and infects sheep (the primary host) when contaminated grass containing the larvae is swallowed.

Triassic: The earliest period of the Mesozoic era. It began about 225 million years age, following the Permian, the last period of the Palaeozoic era, and extended until about 190 million years ago when it was succeeded by the Jurassic It was named, by F. von Alberti in 1834, after the sequence of three, divisions of strata that he studied in central Germany: Bunter, Muschelkalk, and Keuper. The Triassic rocks are frequently difficult to: distinguish from the Permian strata and the term *New Red Sandstone* is often applied to rocks of the Permo-Triassic. During the period marine animals diversified molluscs were the dominant invertebrates ammonites were abundant and bivalves replaced the declining brachiopods. Reptiles were the dominant vertebrates and included turtles, phytosaurs, dinosaurs, and the marine ichthyosaurs.

Tribe: A category used in the classification of plants and animals that consists or several similar or closely related genera within a family. For example the Bambuseae, Oryzeae Paniceae and Aveneae arc tribes of grasses.

Trichome: A hairlike projection from a plant epidermal cell. Examples include root hairs and the stinging hairs of nettle leaves.

Tricuspid valve: A valve, consisting of three flaps, situated between the right atrium and the right ventricle of the mammalian heart. When the right ventricle contracts, forcing blood into the pulmonary artery, the tricuspid valve closes the aperture to the atrium, thereby preventing any backflow of blood. The valve reopens to allow blood to flow front the atrium into the ventricle.

Triglyceride (triacylglycerol): An ester of glycerol (propane 1,2,3-triol) in which all three hydroxyl groups are esterified with a fatty acid.

Triglycerides are the major constituent of fats and oils and provide a concentrated food energy store in living organisms as well as cooking fats and oils, margarines, soaps, etc Their physical and chemical properties depend on the nature of their constituent fatty acids, In simple triglycerides all that fatty acids are identical; in *mixed triglycerides* two or three different fatty acids are present.

Trilobite: An extinct marine arthropod belonging to the class Trilobita (some 4000 species), fossils of which are found in deposits dating front the Procambrian to the Permian period (590280 million years ago). Trilobites were typically small (17 cm long); the oval flattened body comprised a head (covered by a semicircular dorsal shield) and a thorax and abdomen, which were protected by overlapping dorsal plates with a raised central pan and flattened lateral portions, presenting a three lobed appearance. The head bore a pair of antennelike appendages and a pair of compound eyes; nearly all body segments bore a pair of Y-shaped (biramous) appendages: one branch for locomotion and the other fringed for respiratory exchange. Trilobites were bottom dwelling scavengers.

Triploblastic: Describing an animal having a body composed of three embryonic cell layers: the ectoderm, mesoderm, and endoderm. Most multicellular animals are triploblastic; the coelenterates, which are diploblastic, are an exception.

Trochanter: Any of several bony knobs on the femur of vertebrates to which muscles are attached. The second segment of an insect's leg, between the coxa and the femur.

Turbellaria: A class of free-living flatworms comprising the planarians, which occur in wet soils, fresh water, and marine environments. Their undersurface is covered with cilia, used for gliding over stones and weeds. Planarians can also swim by means of undulations of the body.

Turgor: The condition in a plant when its vacuole is distended with water, pushing, the protoplasm against the cell wall. In this condition the osmotic pressure forcing water in is balanced by pressure exerted by the cell wall. Turgidity assists in maintaining the rigidity of plants; a decrease in turgidity leads to wilting.

Turion: A winter bud, covered with scale leaves and mucilage, that is produced by certain aquatic plants, such a, frogbit. Turions become detached and remain dormant on the pond or lake bottom during the winter before developing into new plants the *following season*.

U

Ultracenftifuge: A highspeed centrifuge used te mmure the rate of sedimentation of colloidal particles or te separate macrornoloeules, such as proteins or auclek acids, front solutions, Ultracentrifuges are electrically driven and are capable of speeds up te 60 000 rpm.

Ultramicroscope: A form of microscope that reveals the prtwiice, of particles that cannot bc seen with a norniai optical micrscope. Colloidal partiuke, smoke particles, etc., are suspended in a liquid or gas in a cell with a black background arnd illuminated by an intense cone of light that enters the cell from the side wise its apex in the field of view. The partielles then produce diffractionring systems, appeafing as brigtit specks on the dark background

Ultrasonics: The study and use of pressure waves that have a frequency m excess of 20000 Hz and art therefore inaudible te the human ear. Ultrasound is used in medical diagnosis, particularly in conditions such as pregnancy, in which Xrays could have a harinful effect.

Ultrastructure: The submicroscopic, almost molecular, structure of living cells, which is revealed by the use of an electron rnicroscope

Ultraviolet radiation (UV): Electromagnetic radiation having wavelengths betwoen that of violet light and long X-rays, between 400 nanometres and 4 am. In the range 400-300 nm the radiation is known as the near ultraviolet, In the range 300-200 uni it is known as the far ultraviolet. Below 200 nm it is known as the extreme ultraviolet or the vacuum ultraviolet, as absorption by the oxygen in the air makes the use of evacuated apparatus essenfial. Most UV radiation for pracucation is produced by varions types of the uryvapour lamps. Ordinary g1w absort UV radiation and therefore lenses and prisins for use in the UV are made from quartz.

Umbel: A type of racemose inflorescence in which stalked flowers arise from the saine point on the flower axis, reembling the spokes of an umbrella. An involucre (cluster) of bracts may coeur ai the point where the stalh emerge. This arrangement is characteristic of the family Umbellferae, in which the inflorescence is usually a compound urnbel.

Umbilical cord: The cord that connects the embryo te the placenta in mammals. It contams a vent and two artenes that carry blood betwoen the embryo and placenta It is severed after birth of the newly borit animal from the placenta, and slirivels te leave a scar, the navet, on the animal.

Ungulate: A herbivoreus marnmal with hoofed feet unguligrade). Ungulates are grouped into orders Artiodactyla and Perisodactyla.

Unicellular: Describing tissues, organs, or organisms consisting of a single cell. For example, the reproductive organs of sente allgae and fungilare unicellular. Unicel-ular organisms, e.g. bacteria, protozoans, and certain algac and fungi, are sometimes placed in a separate kingdom, the Protista.

Unisexual: Describing animals or plants with cither male or female reproductive organs but not both. Most of the more advanced animals are unisexual but plants are often hermaphrodite. Flowers that contain either stamens or carpels but not both are also described as unisexual.

Unit: A specified measure of a physical quantity, such as length, rnass, time, etc., specified multiples of which are used to express magnitudes of that physical quantity. For scientific purposes previous systems of units have now been replaced by SI units.

Uracil: A pyrimidine derivative and one of the major component bases of m_cleotides and the nucleic acid RNA.

Urea (carbamide): A white crystalline watersoluble sofid, $CO(NH_2)$ Urea is the major end product of nitrogen excretion in maminals, being synthesized by the urea cycle. Urea is synthesized industrially from ammonia and carbon dioxide for use in ureaformaldehyde resins and pharmaceuticals, as a source of nonprotein nitrogen for ruminant livestock, and as a nitrogen fertilder.

Urea cycle (ornithine cycle): The series of biochemical reactions that couverts ammonia to urea during the excretion of metabolic nitrogen. Urea formation occurs in manimals and, to a lesser extent, in some other animals. The liver couverts ammonia to the much less toxic urea, which is excreted in solution in urine.

Urethm: The duct in mammals that conveys urine from the bladder to be discharged to the outside of the body. In males the tirethra passes through the pems and is joined by the vas defrens; it therefore also serves as a channel for sperm.

Uric acid: The end product of purine breakdown in most primates, birds, terrestrial reptiles, and insects and also, (except in primates) the major form in which metabolic nitrogen is excreted. Being fairly insoluble, uric acid can be expelled in solid form, which conserves valuable water in arid envimoments, The accumulation of usic acid in the synovial fluid of joints causes gout.

Urine: The aqueous fluid formed by the excretory organs of animais for the removal of metabolic waste producis. In higher animals, urine is produced by the kidneys, stored in the bladder, and excreted through the 'urethra or cloaca. Apart from water, the major constituents of urine are one or more if the end products of nitrogen metabolism armuoqia, urea, uric acid, and creatinine. It may also contain various inorganic ions, the pigments urochrome and urobilin, amino acids, and purines. Precise composition depends on man factors, especially the habitat of a particular species: aquatic ammals produce copious vo!umes ierrestrial animals need to consme water and produce much less.

Uterus (womb): The orgin of female mainmals in which the embryo develops. Paired in mess manimals but single in humans, it is situated between the bladder and rectum and is cormected to the fallopian tubes and to the agina. Cyclical changes are associated with egg production and provides a thick spongy layer in which the fertilized egg becomes embedded. The outer wall of the uterus is thick and muscular; by contracting, it forces the fully grown fetus through the vagina to the outside.

Vaccine: A liquid preparation of treated disease-producing microorganisms or their products used to stimulate an immune response in the body and so confer resistance to the disease. Vaccines are administered orally or by injection (*inoculation*). They take the form of dead uses or bacteria that can still act as antigens, live but weakened microorganisms, specially treated toxins, or antigenic extracts of the microorganism.

Vacuole: A space within the cytoplasm of a living cell that is filled with air, water or other liquid, sap, or food particles. In plant cells there is usually one large vacuole animal cells usually have several small one.

Vagina: The tube leading from the uterus to the outside. Sperm are deposited in the vagina during copulation and the fully developed fetus is born through it. In a number of mammals the vagina may be sealed when the animal is not sexually receptive and only open during oestrus. Its lifting produces mucus, which prevents friction and the entry of infective organisms.

Vagus nerve: The tenth cranial nerve; a paired nerve that supplies branches to many major internal organs. It carries motor nerve fibres to the heart, lungs, and viscera and sensory fibres from the viscera.

Variation: The deferences between individuals of a plant or animal species. Variation may be the result of environment conditions; for example, water supply and light intensity affect the height and leaf size of a plant Differences of this kind, acquired during the lifetime of an individual, are not transmitted in succeeding generations since the genes are not affected. Variations due to difference in genetic was constitution are inherited. Wide genetic variation improves the ability of a species to survive in a changing environment, since the chances that some individuals which tolerate a particular change are increased.

Such individuals will survive and transmit the advantageous genes to their offspring.

Variety: A catch used in the classification of plants and animals below the species level. A variety consists of a group of individuals that differ distinctly from but can interbreed with other varieties of the same species. The characteristics of a variety are genetically inherited. Examples of varieties include breeds of domestic animals and the races of man.

Vascular bundle (fascicle): A long continuous strand of conducting (vascular) tissue in seed plants and pteridophytes that extends from the roots through the stem and into the leaves. It consists of xylem and phloem, which are separated by a cambium in plants that undergo secondary thickening.

Vascular plants: All plants possessing organized vascular tissue. They may be grouped in a single division, the Tracheophyta, or separated into two divisions, the Pteridophyta and Spermatophyta.

Vascular system: 1. A specialized network of vessels for the circulation of fluids throughout the body tissues of an animal. All animals, apart from protozoans and other simple invertebrate groups, possess a: *vascular system,* which enables the passage of respiratory gases, nutrients, excretory products, and other metabolites into and out of the cells. In vertebrates it consists of a muscular heart, which pumps blood through major blond vessels (arteries) into increasingly liner branches until in the capillaries it is in intimate contact with tissues. It then returns to the heart via another network of vessels (the veins). This circulation also enables a stable internal environment for tissue function, the transmission of chemical messengers (hormones) around the body, and a means of defending the body against pathogens and damage via the immune system. A water vascular system is characteristic of the Echinodermata. 2. The system of vascular tissue in plants.

Vascular tissue (vascular system): The tissue that conducts water and nutrients through the plant body in higher plants (ferns and seed plants). It consists of xylem and phloem. Since the xylem and phloem tissues are always in close proximity to each other, distinct regions of vascular tissue can be identified. The possession of vascular tissue base enabled the higher plants to attain a considerable size and dominate most terrestrial habitats.

Vas deferens: One of a pair of ducts carrying sperm from the testis to the outside, in mammals through the urethra.

Vas efferens: Any of various small ducts carrying sperm. In reptiles, birds, and mammals they convey sperm from the *seminiferous tubules* of the testis to the epididymis; in invertebrates they carry sperm, from the testis to the vas deferens.

Vasoconstriction: The reduction in the internal diameter of blood vessels, especially arteriole; or capillaries. The constriction of arterioles is mediated by the action of nerves on the smooth muscle fibres of the arteriole walls and results in an increase in blood pressure.

Vasodilation (vasodilatation): The increase in the internal diameter of blood vessels, especially arterioles or capillaries. The vasodilation of arterioles is mediated by the action of nerves on the smooth muscle fibres of the arteriole walls and results in a decrease in blond pressure.

Vasomotor nerves: The nerves of the autonomic nervous system that control the diameter of blond vessels. Vasoconstrictor nerves decrease the diameter; *vasodilator* nerves increase it.

Vasopressin (antidiuretic hormone; ADH): A hormone, secreted by the posterior pituitary gland, that stimulates absorption of water by the kidneys and thus controls the concentration of body fluids. Vasopressin is produced by specialized germ cells in the hypothalamus of the brain and is transported to the posterior pituitary in the bloodstream. Deficiency of ADH results in a disorder known as diabetes insipidus, in which large volumes of urine are excreted; it is treated by administration of natural or synthetic hormone.

Vector: An animal, usually an insect, that passively transmits disease-causing micro-organisms from one animal or plant to another or from an animal to man.

Vegetative propagation (vegetative reproduction): A form of asexual reproduction in plants whereby new individuals develop front specialized multicellular structures (e.g. tubers, bulbs) that become detached from the parent plant. Examples are the production of strawberry plants front runners and of gladioli front daughter corms. Artificial methods of vegetative propagation include grafting, budding and making cuttings. Asexual reproduction in animals, e.g. budding in Hydra.

Vein: A blood vessel that carries blood towards the heart. Most veins carry deoxygenated blond (the pulmonary vein is an exception). The largest veins six fed by smaller ones, which are formed by the merger of venules. Veins have thin walls and a relatively large internal diameter. Valves within the veins ensure that the flow of blond is always towards the heart. A vascular bundle in a leaf. Any of the tubes of chitin that strengthen an insect's wing.

Vena cava: Either of the two large veins that carry deoxygenated blond into the right atrium of the heart. The precaval vein (anterior or superior vena cava) receives blond from the head and forelimbs; the postcaval vein drains blood from the trunk and hindlimbs.

Venation: The arrangement of veins (vascular bundles) in a leaf. The leaves of dicotyledons have a central main vein (midrib) with side branches that themselves further subdivide to form a network (*net* or *reticulate venation).* The leaves of monocotyledons have parallel veins *(parallel venation).* 2. The arrangement of the veins in an insect's wing, which is often important in classification.

Ventral: Describing the surface of a plant or animal that is nearest or next to the ground or other support, i.e. the lower surface. In bipedal animals, such as man, it is the forward directed (anterior) surface.

Ventricle: A chamber of the heart that receives blood from an atrium and pumps it into the arterial system. Amphibians and fish have a single ventricle, but mammals, birds, and reptiles have two, pumping deoxygenated blood to the lungs and oxygenated blood to the test of the body, respectively. 1 Any of the four linked fluid-filled cavities in the brain of vertebrates. One of these cavities is in the medulla oblongata, two are in the cerebral hemispheres, and the fourth is in the posterior part of the forebrain.

Vernalization: The application of a cold treatment to germinating seeds or seedlings to, ensure flowering. Many plants, including winter cereals, will not flower unless subjected to a period of chilling early in their development. Winter cereals are therefore sown in the autumn for flowering the following year. However, if germinating seeds are artificially vernalized they can be sown in the spring for flowering the same year.

Vertebra: Any of the bones that make up the vertebral column. In mammals each vertebra typically consists of a main body, or *centrum,* from which arises a *neural arch* through which the spinal cord passes, and *transverse processes* projecting from the side There are five groups of vertebrae, specialized for various functions and varying in number with the species. In man, for example, there are 7 cervical vertebrae, 12 thoracic vertebrae, 5 lumbar vertebrae, 5 fused sacral vertebrae, and 5 fused caudal vertebrae (forming the coccyx).

Vertebral column (backbone; spinal column; spine): A flexible bony column in vertebrates that extends down the long axis of the body and provides the main skeletal support. It also encloses and protects the spinal cord and provides attachment for the muscles of the back. The vertebral column consists of a series of bones separated by discs of cartilage *(intervertebral discs),* It articulates with the skull by means of the atlas vertebra, with the ribs at the thoracic vertebrae, and with the pelvic girdle at the sacrum.

Vertebrata (Craniata): The largest subphylum of the Chordata, comprising ail those animals with backbones. The replacement of the rigid notochord by the more flexible backbone has permitted vertebrates a greater degree of movement and subsequent improvement in the sense organs and enlargement of the brain. The brain is enclosed in a skeletal case, the cranium. There are seven extant classes: Agnatha (jawless fishes), Elasmobranchii (cartilaginous fishes), Osteichthyes (bony fishes), Amphibia, Reptilia, Aves (birds), and Mammalia.

Vesicle: A sine usually fluid-filled, membrane bound sac within the cytoplasm of a living cell. Vesicles form part of the Golgi apparatus.

Vessel (in botany): A tube within the xylem composed of joined vessel element Vessels facilitate the efficient movement of water front the roots to the shoots and leaves of a plant, 1 (in zoology) Any of various tubular structures through which substances are transported, especially a blood vessel or a lymphatic vessel.

Vessel element: A type of cell occurring within the xylem of flowering plants, many of which, end to end, form water-conducting vessels. Vessel elements arc frequently very broad and have side walls thickened by deposits of lignin over most of the surface area. However the end walls are broken down to provide connections with the cells both above and below them.

Vestigial organ: Any part of an organism that has diminished in size during its evolution because the function it served decreased, in importance or became totally unnecessary.

Vibrio: Any comma-shaped bacterium. Generally, vibrios are Gram-negative, motile, and aerobic. They are widely distributed in soil and water and while most feed on dead organic matter same are parasitic, e.g. Vibrio cholerae, the causal agent of cholera.

Villus: A microscopic outgrowth from the surface of sortie tissues and organs, which serves to increase the surface area of the organ. Numerous villi line the interior of the small intestine. Their shape may vary from fingerlike (in the duodenum) to spadelike (in the ileum). Intestinal villi are specialized for the absorption of soluble food material: each contains blood vessels and a lymph vessel.

Villi also occur on the chorion of the mammalian placenta, where they increase the surface area for the exchange of materials between the fetal and maternal blood.

Virulence: The disease producing ability of a microorganism.

Virus: A particle that is too small to be seen with a light microscope or to be trapped by fitters but is capable of independent metabolism and reproduction within a living cell. Outside its host cell a virus is completely inert. A mature virus (a *virion*) ranges in size train 20 to 400 am in diameter. It consists of a core of nucleic acid (DNA or RNA) surrounded by a protein coat (*capsid*). Same bear an outer envelope (*enveloped viruses*). Inside its host cell the virus initiates the synthesis of viral proteins and undergoes replication. The new virions are released when the host cell disintegrates. Viruses are parasites of animals, plants, and same bacteria. Viral diseases of animals include the common cold, influenza, smallpox, herpes, hepatitis, polio, and rabies; some viruses are al o implicated in the development of cancer. Plant viral diseases include various forms of yellowing and blistering of leaves and stem. Antibiotics are ineffective against viral diseases but vaccines provide good protection.

Vital staining: A technique in which a harmless dye is used to stain living tissue for microscopical observation. The stain may be injected into a living animal and the stained tissue removed and examined or the living tissue may be removed directly and subsequently stained. Microscopic organisms, such as Protozoa, may be completely immersed in the dye solution. Vital stains include trypan blue, vital red and Janus green, the latter being especially suitable for observing mitochondria.

Vitamin: One of a number of organic compounds required by living organisms in relatively small amounts to maintain normal health. There are some 14 generally recognized major vitamins: the water-soluble vitamin B complex (containing 9) and vitamin C and the fat-soluble vitamin A, vitamin D, vitamin E, and vitamin K. Most B vitamins and vitamin C occur in plants, animals, and microorganisms;

they function typically as coenzymes. Vitamins A, D, E, and K occur only in animals, especially vertebrates, and perform a variety of metabolic roles. Animals are unable to manufacture many vitamins themselves and must have adequate amounts in the diet. Foods may contain vitamin precursors (called *provitamins)* that are chemically changed in the actual vitamin on entering the body. Many vitamins are destroyed by hight and heat, e.g., during cooking.

Vitamin A (retinol): A fat-soluble vitamin that cannot be synthesized by Mammals and ether vertebrates and must be provided in the diet. Green plants contain precursors of the vitamin, notably carotenes, that are converted to vitamin A in the intestinal wall and liver. The aldehyde derivative of vitamin A, *retinal, is* a constituent of the visual pigment rhodopsin. Deficiency affects the eyes, causing night blindness, xerophthalmia, and eventually total blindness. The role of vitamin A in other aspects of metabolism is less clear; it may be involved in controlling ATP production and the growth of epithelial cells.

Vitamin B complex: A group of water-soluble vitamins that characteristically serye as components of coenzymes. Plants and many micro-organisms can manufacture B vitamins but dietary sources are essential for most animals. Heat and light tend to, destroy B vitamins

Vitamin B_1 (thiamine) is a precursor of the coenzyme thiamine pyrophosphate, which functions in carbohydrate metabolism. Deficiency leads to beriberi in humans and to polyneuritis in birds. Good sources include brewers yeast, wheatgerm, beans, peas, and green vegetables.

Vitamin B_2: (*riboflavin*) occurs in green vegetables, yeast, liver, and milk. It is a constituent of the coenzymes FAD and FMN, which have an important role in the metabolism of ail major nutrients as well as in the oxidative phosphorylation reactions of the electron transport chain. Deficiency of B, causes inflammation of the tongue and lips and mouth sores.

Vitamin B_6 (pyridoxine) is widely distributed in cereal grains, yeast, liver, milk, etc., it is a constituent of a coenzyme (pyridoxal phosphate) involved in amino acid metabolism. Deficiency causes retarded growth, dermatitis, convulsions, and other symptoms.

Vitamin B_{12} (*cyanocobalamin*) is manufactured only by microorganisms and natural sources are entirely of animal origin. Liver is especially rich in it. One form of B_{12} functions as a coenzyme in a number of reactions including the oxidation of fatty acids and the synthesis of DNA. It also works in conjunction with folic acid (another B vitamin) in the synthesis of the amino acid methionine and it is required for normal production of red blood cells. Vitamin B_{12} can only be absorbed from the gut in the presence of a glycoprotein called intrinsic factor; lack of this factor or deficiency of B_{12} results in pernicious anaemia.

Other vitamins in the B complex include nicotinic acid, pantothenic acid, folic acid, biotin, and lipoic acid.

Vitamin C (ascorbic acid): A colourless crystalline water-soluble vitamin found especially in citrus fruits and green vegetables. Most organisms synthesize it from glucose but man and other primates and various other species must obtain it from their diet. It is required for the maintenance of healthy connective tissue; deficiency leads to scurvy. Vitamin C is readily destroyed by heat and light.

Vitamin D: A fat-soluble vitamin occurring in the form of two steroid derivatives: vitamin D_3 found in yeast and vitamin D_2, which occurs in animals, Vitamin D_2 is formed front a steroid by the action of ultraviolet light and D_2 is produced by the action of sunlight on a cholesterol derivative in the skin, Fish-liver oils are the major dietary source. The active form of vitamin D is manufactured in response to the secretion of parathyroid hormone, which occur when blood calcium levels are low. It causes increased uptake of calcium from the gut which increases the supply of calcium for bone synthesis. Vitamin D deficiency causes rickets in growing animals and osteomalacia in mature animals.

Vitamin E: (tocopherol) A fat-soluble vitamin consisting of several closely related compounds, deficiency of which leads to a range of disorders in different species, including muscular dystrophy, liver damage, and infertility. Good sources are cereal grains and green vegetables.

Vitamin K: A fat-soluble vitamin consisting of several related compounds that act as coenzymes in the synthesis of several proteins necessary for blood clotting. Deficiency of vitamin K, which leads to extensive bleeding, is rare because a form of the vitamin is manufactured by intestinal bacteria. Green vegetables and egg yolk are good sources.

Vitreous humour: The colourless Jelly that fills the space between the lens and the retina of the vertebrate eye.

Viviparity: (in zoology) A form reproduction in animals in which the developing: embryo obtains its nourishment directly from the mother via a placenta or by other means. Viviparity occurs in some insects and other arthropods, in certain fishes, amphibians, and reptiles, and in the majori'y of mammals. 2. (in botany) a. A form of asexual reproduction in certain plants, such as the onion, in which the flower develops into a budlike structure that forms a new plant when detached from the parent. b. The development of young plants on the inflorescence of the parent plant, as seen in certain grasses and the spider plant.

Vocal cords: A pair of elastic membranes that project into the larynx in airbreathing vertebrates. Vocal sounds are produced when expelled air passing through the larynx vibrate, the cords. The pitch of the sound produced depends on the tension of the cords, which is controlled by muscles and cartilages in the larynx.

Voluntary muscle (skeletal, striped, or striated muscle): Muscle that is under the control of the will and is generally attached to the skeleton. An individual muscle consists of bundles of long *muscle fibres*, each containing many nuclei, the whole muscle being covered with a strong connective tissue sheath and attached at each end to a bone by inextensible tendons. Each fibre contains smaller fibres (*myofibrils*) having alternate light and dark bands (*sarcomeres*), which are responsible for the muscle's contractile ability and its typical striped appearance under the microscope. The end of the muscle that is attached to a nonmoving bone is called the origin of the muscle; the end attached to a moving bone is the insertion. As a muscle contracts it becomes shorter and fatter, moving one bone closer to the other.

Wallace's line: An imaginary line that runs between the Indonesian islands of Bali and Lombok and represents the separation of the Australian and Oriental faunas. It was proposed by the naturalist A. R. Wallace who had noted that the mammals in Southeast Asia are different from and more advanced than their Australian counterparts. He suggested this was because the Australian continent had split away front Asia before the better adapted placental mammals evolved in Asia. Hence the isolated Australian marsupials and monotremes were able to thrive while those in Asia were driven to extinction by competition from placental mammals.

Warfarin: 3-(alpha-acetonyl-benzyl)-4-hydroxycoumarin: a synthetic anticoagulant used both therapeutically in clinical medicine and, in lethal doses, as a rodenticide.

Warning coloration (aposematic coloration): The conspicuous markings of an animal that make it easily recognizable and warn would be predators that it is a poisonous, foul-tasting, or dangerous species.

Watson-Crick model: The double-stranded twisted ladder-like molecular structure of DNA as determined by the US biochemist James Watson and the British biochemist Francis Crick at Cambridge, England, in 1953 It is commonly known as the *double helix.*

Wax: Any of various solid or semisolid substances. There are two main types. Mineral waxes are mixtures of hydrocarbons with high molecular weights. Waxes secreted by plants or animals are mainly esters of fatty acids and usually have a protective function. Examples are the beeswax forming part of a honeycomb and the wax coating on sortie leaves, fruits, and seed coats, which acts as a protective water-impermeable layer supplementing the functions of the cuticle. The seeds of a few plants contain wax as a food reserve.

Weismannism: The theory of the *continuity of the germ plasm* published by August Weismann in 1886. It proposes that the contents of the reproductive cells (sperms and ova) are passed on unchanged from one generation to the next, unaffected by any changes undergone by the rest of the body. It thus rules out any possibility of the inheritance of acquired characteristics, and has become fundamental to neo-Darwinian theory.

Whalebone (baleen): Transverse horny plates hanging down front the upper jaw on each side of the mouth of the toothless whales, forming a sieve. Water, containing plankton on which the whale feeds, enters the open mouth and is then expelled with the mouth slightly closed, so that food is retained on the baleen plates.

White matter: Part of the tissue that makes up the central nervous system of vertebrates. It consists chiefly of nerve fibres enclosed in whitish myelin sheaths.

Wild type: Describing the form of an allele possessed by most members of a population in their natural environment. Wild-type alleles are usually dominant.

Wood: The hard structural and water-conducting tissue that is found in many perennial plants and forms the bulk of trees and shrubs.

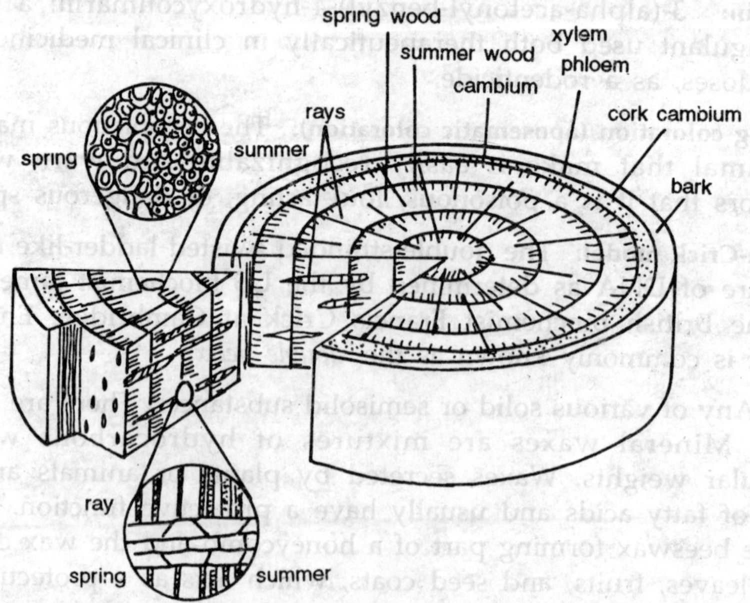

Diagramatic section through a woody stem with four season's growth

It is composed of secondary xylem and associated cells, such as fibres. The wood of angiosperms is termed *hardwood,* e.g. oak and mahogany, and that of gymnosperms *softwood,* e.g. pine and fir. New wood is added to the outside of the old Wood each growing season by divisions of the vascular cambium. Only the outermost new wood functions in water conduction; the inner wood provides only structural support.

Woody perennial: Trees and shrubs persist above ground throughout the year but may show adaptations to survive unfavourable seasons.

X

Xeromorphic: Describing the structural modifications of certain plants (xerophytes) that enable them to reduce water loss, particularly from their leaves and stems.

Xerophyte: A plant that is adapted to live in conditions in which there is either a scar city of *water* in the soil or the atmosphere is dry enough to, provoke excessive transpiration, or both. Xerophytes have special structural *(xeromorphic)* and functional modifications, including swollen water-storing stems and specialized leaves that may be hairy, rolled, or reduced to spines or have a thick cuticle to, lower the rate of transpiration.

X-ray crystallography: The use of X-ray diffraction to, determine the structure of crystals or molecules, such as nucleic acids. The technique involves directing a beam of X-rays at a crystalline sample and recording the diffracted X-rays on a photographic plate. The diffraction pattern consists of a pattern of spots on the plate, and the crystal structure can be worked out front the positions and intensities of the diffraction spots. X-rays are diffracted by the electrons in the molecules and if molecular crystals of a compound are used, the electron density distribution in the molecule can be determined.

X-rays: Electromagnetic radiation of shorter wavelength than ultraviolet radiation and longer wavelength than gamma radiation. The range of wavelengths is 10^{-11} m to 10^{-9} m. X-rays can pass through many forms of matter and they are therefore used medically and industrially to examine internal structures. X-rays are produced for these purposes by an X-ray tube.

Xylem: A tissue that transports water and dissolved mineral nutrients in vascular plants. In flowering plants it consists of hollow vessels that are formed front cells joined end to end. The end *walls* of the vessel elements are performated to allow the passage of water. In less advanced vascular plants, such as conifers and ferns, the constituent cells of the xylem are called tracheids. In young plants and at the shoot and root tips of older plants the xylem is formed by the apical meristems. In plants showing secondary growth this xylem is replaced in most of the plant by secondary xylem, formed by the vascular cambium.

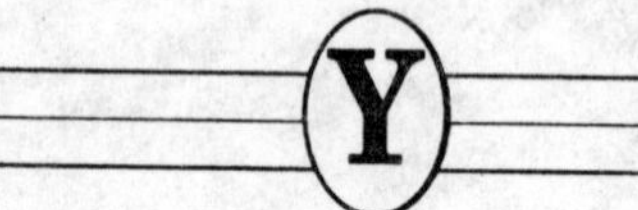

Y chromosome: The smaller of the two types of sex chromosome, found only in the heterogemametic sex.

Yeasts: A group of unicellular fungi many of which belonged to the Ascomycetes. Certain species of the genus Saccharomyces are used in the baking and brewing industries.

Yolk: The food stored in an egg for the use of the embryo. It can consist mainly of protein (protein yolk) or of phospholipids and fats. The eggs of oviparous animals (eg. birds) contain a relatively large yolk.

Z

Zona pellucida: The thick clear membrane surrounding the mammalian egg. It is surrounded by cumulus cells in the freshly ovulated egg, but these disperse as the sperm pass between them and penetrate the zona by enzyme action.

Zoogeography: The study of the geographical distributions of animals. The earth can be divided into several zoogeographical regions separated by natural harriers, such as occans, deserts, and mountain ranges. The characteristics of the fauna of each region are believed to depend particularly, on the process of continental drift and the stage of evolution reached when the various land masses became isolated.

Zoology: The scientific study of animals, including their anatomy, physiology, biochemistry, genetics, ecology, evolution, and behaviour.

Zooplankton: The animal component of plankton. All major animal phyla are represented in zooplankton, as adults, larvae, or eggs; some are just visible to the naked eye but most cannot be seen without magnification. Near the surface of the sea there may be many thousands of such animals per cubic metre.

Zygospore: A zygote with a thick resistant wall formed by some algae and fungi. It results from the fusion of two gametes, neither of which is retained by the parent in any specialized sex organ. It enters a resting phase before germination.

Zygote: A fertilized female gamete: the product of the fusion of the nucleus of the ovum or ovule with the nucleus of the sperm or pollen grain.

Zygote: The diploid cell resulting from the fusion of two haploid gametes. A zygote usually undergoes cleavage immediately.